Extended Families of Generalized Tukey-Lambda Distribution and Their Applications

Tukey-Lambda Distribution and Their Applications

Extended Families of Generalized Tukey-Lambda Distribution and Their Applications

Tukey-Lambda Distribution and Their Applications

Bachioua Lahcene

ELIVA PRESS

ISBN: 978-99949-8-060-4

Cover Design: Eliva Press
Cover Image: Freepik Premium
Email: info@elivapress.com
Website: www.elivapress.com

This Eliva Press imprint is published by the registered company Eliva Press Global Ltd. part of Eliva Press S.R.L. Publishing Group

The registered company address is: Pope Hennessy Street Level 2, Hennessy Tower Port Louis, Mauritius
Eliva Press S.R.L. Publishing Group address is: Bulevardul Moscova 21, Chisinau, Moldova, Europe

Preface This book arose from our experience as active researchers working in the field of probability distributions and have derived classes, extended families of Generalized Tukey Lambda Distribution, in our research during the past two decades. We have noticed that it was appropriate to put forward a set of choices and ideas that were used after the forties of the past century and before the emergence of influential probability distributions, which paved the way for their current enormous development. In particular, we indicate to the family of quantile that generalizes its basic form. In addition to the extension of extended families of Generalized Tukey Lambda Distribution model and its various generalizations.

The astonishing computational ability of personal computing is now making cutting edge inferential technologies seamlessly accessible. Although the computing power available is more than equal to the task, many powerful techniques for extracting the most information from expensive data are not as widely adopted as they should be. due to unavailability of appropriate software.

We tried to write this book that we wanted to be out five years ago, and we began to organize the material for research and review the most important research ideas on the topic at the level of statistics, quality control, reliability engineering, fitting and simulation, among others. We have considered the necessary license for using published material in order to include a lot of material in the current form of the book

This book facilitates the basic elements of using a computer to determine steps and methods to make necessary performing Monte Carlo quantifications for practical inference. We hope this book will help researchers and graduate students who are interested in dealing with flexible, general and extended forms of probability distributions. Also, it provides other quantitative stochastic models that perform the necessary simulations for inference in single and multiple samples.

The book considers, as well, the special cases of the generalized form of the extended gamma distribution, and other related systems of probability distributions. All chapters of this book conclude with a set of exercises that can be used to help the readers to understand the material in depth or as class exercises in a lecture or short course. The families of probability distributions cannot be discussed in isolation from the rest of the statistical methodology.

Knowledge of elementary probability theory and distributions such as the Pearson system, Burr system, Johnson system, Tukey lambda distribution system, exponential distribution, normal distribution, and natural logarithm are essential for smooth progress through the book and effective acquaintance of its application. The book aims to be self-sufficient in this regard, with the inclusion of two chapters, at the beginning, dealing with the basics of probability and the necessary statistics at the beginning of the book. The sequel of the goes as follows:

Chapter 1 introduces the ideas of Chapter One General Families of Quantile Transform Distributions , probability density and cumulative Distribution function by mutual exclusivity, independence, conditional probability, and the Law of total probability as they impact application - related calculations. The chapter concludes with applications of probability principles to the computation of the probability of systems in terms of the distribution of its components.

chapter 2 separately and jointly describes the origin & history of Tukey lambda distribution and general families of quantile transform distributions and same applications of Tukey lambda distribution with exercises in probability theory, the inequalities of continuous random variables, and the concepts of mean, variance and covariance. Known random variables that might arise sometime later, are included.

chapter 3 lists the characteristics of an expanded generalized Tukey lambda distribution and measurement of skewness and kurtosis for the Tukey lambda distribution. it is shown to be a generalization of the shape characteristics of the FMKL parameterization .

It gives expressions for the mean, variance, mode, skew, risk function, and quantities for a normal distribution in terms of the two parameters. it is shown that the transformation of the random variable follows the distribution of the smallest extremes. the shape and form transformation of the random variable for the various sub-states is shown from the general formula.

chapter 4 describes a number of useful applications of the on extended Tukey's lambda distribution models with satisfy introduction and new proposed model and application in percentile function of some known distribution with quantitative techniques.

Chapter 5 treats extended Some Application of Extended Tukey's Lambda Distribution Models . The notions of bias and precision of estimation are illustrated by means of the sampling distributions of two competing estimators of the 10th percentile of a population.

Graphic large al estimation, which for years was the standard means of estimating the Weibull parameters, is explained and various choices of plotting positions are described and compared. presents general forms extended of Tukey Lambda model. It also discusses their: properties, estimation and some of their applications.

Appropriate statistical distributions help answer these questions. Focusing on techniques used successfully in many areas, the author presents all relevant results related to generalized Lambda distribution (EGLD), generalized Bootstrap (GB), and Monte Carlo (MC) simulation.

They provide the tables, algorithms, and computer programs needed to fit continuous probability distributions of data in a variety of conditions covering bivariate as well as univariate distributions, including situations where there are no moments.

Regardless of your specific field of statistical science, social science, or statistics, statistical distributions for the appropriate practitioner or theorist are required reading.

It includes extensive applications illustrating methods in practice and provides evidence of key findings for those involved in theoretical development. Without it, you may be using outdated methods, wasting time, and risking incorrect results.

Table of Contents

CHAPTER ONE:

GENERAL FAMILIES OF QUANTILE TRANSFORM DISTRIBUTIONS

1.1.ORIGIN & HISTORY

Historically, the Pearson and Johnson families of distributions have been used when shape flexibility is needed. That is because both families can match the first four moments (mean, variance, skewness, and kurtosis) of any data set. In many cases, however, these distributions are either difficult to fit to data or not flexible enough to fit the data appropriately.

Quintiles have a long history in applied statistics, especially the median. The analysis of astronomical data by Galileo Galilei in 1632 and geodic measurements by Roger Boscovich in 1757 are presumably the earliest examples of application of the least absolute deviation (L1L1) estimator in its, respectively, unconditional and conditional forms. The theoretical studies on quintiles of continuous random variables started to appear in the statistical literature of the 20th century.

Quantile and conditional quantile statistical methods are not widely practiced in introductory statistics courses. They were pioneered by Galton , who computed medians and quartiles of conditional distributions of heights of sons given heights of parents, and discovered that they had constant scale and linear location. Galton thus pioneered regression, correlation, vicariate normal distributions and conditional normal distributions.

Many facts about quintiles have a long history and were known before 1900 . Quantile statistical thinking, as I have innovated it in my research since Parzen , is outlined in this paper. My teaching philosophy has as its maxim: to earn more, learn more, and believe that learning a lot (answering all related questions) is easier than learning little (answering only the questions asked).

In the case of discrete data, studies have somewhat lagged behind most probably because of the analytical drawbacks surrounding the discontinuities that characterize discrete quantile functions. Some forms of approximation to continuity have been recently proposed to study the large sample behavior of quantile estimators. For example, Ma, Genton, and Parzen have demonstrated the asymptotic normality of unconditional sample quintiles based on the definition of the mid-distribution function. Machado and Santos Silva proposed inferential approaches to the estimation of conditional quintiles for counts based on data jittering.

Here we discuss several distributional families relevant to loss modeling in insurance which can only be specified via the transformation of another standard random variable, for example a Gaussian. Examples of such models which are typically defined through their quantile functions include the Johnson family, with base distribution given by Gaussian or logistic, and the Tukey family with base distribution typically given by a Gaussian or logistic.

The concept of constructing skewed and heavy-tailed distributions through the use of a transformation of a Gaussian random variable was originally proposed in the work of Modern techniques in data analysis and is therefore aptly named the family of Tukey distributions. This family of distributions was then extended by A class of distributions which includes the normal ones. The multivariate versions of these models have been discussed in multivariate g-and-h distribution.

The development of quantile-parameterized distributions was inspired by the practical need for flexible continuous probability distributions that are easy to fit to data. For example, the beta distribution is a flexible Pearson distribution that is frequently used to model percentages of a population. However, if the characteristics of this population are such that the desired cumulative distribution function should run through certain specific commutative distribution function points, there may be no beta distribution that meets this need. Because the beta distribution has only two shape parameters, it cannot, in general, match even three specified cumulative distribution function points. Moreover, the beta parameters that best fit such data can be found only by nonlinear iterative methods.

Practitioners of decision analysis, needing distributions easily parameterized by three because such points were specified as the result of an expert-elicitation process), originally invented quantile-parameterized distributions for this purpose. Keelin and Powley provided the original definition. Subsequently, Keelin developed the metalog distributions, a family of quantile-parameterized distributions that has virtually unlimited shape flexibility, simple equations, and closed-form moments.

Within this family of distributions, two particular subfamilies have received the most attention in the literature; these correspond to the g-and-h and the g-and-k distributions. The first of these families, the g-and-h, has been studied in several contexts. See for instance the developments in the areas of risk and insurance modeling in The quantitative modeling of operational risk, and the detailed discussion in Fundamental Aspects of Operational Risk and Insurance Analytics. The second family of g-and-k models has been looked at in works such as Robustness of ranking and selection rules using generalized g-and-k distributions.

A key advantage of models such as the g-and-h family for modeling losses in a non-life insurance setting is the fact that they provide a very flexible range of skewness, kurtosis, and heavy-tailed features while also being specified as a rather simple transformation of standard Gaussian random variates, making simulation under such models efficient and simple.

It is important to note that the support of the g-and-h density includes the entire real line. In some cases this is appropriate in non-life insurance modeling such as under logarithmic transforms of the loss or claim amounts.

In other cases it may be more appropriate to consider loss models with strictly positive supports, in general this can be achieved either by truncation or by restriction of the parameter values. In some subfamily members, the g-and-h family automatically takes a positive support such as the double h-h subfamily.

Statistical distributions play an important role in theory and applications, which are used to fit model and describe real world phenomena. For this reason, statistical distributions and their properties are of great importance especially in the social sciences (such as economics, political science) and engineering disciplines such as computer science, as well as in the natural sciences (such as biology, chemistry, physics).

Although a large number of distributions have been defined and studied over the years, seeking for more flexibility in fitting data remains a strong reason for researchers to develop and study new distributions.

1.2. RANDOM EXPERIMENT MODELED BY A PROBABILITY SPACE

As usual, our starting point is a random experiment modeled by a with probability space (Ω,F,P). So to review, Ω is the set of outcomes, **F** is the collection of events, and P is the probability measure on the sample space (Ω,F). In this section, we will study two types of functions that can be used to specify the distribution of a real-valued random variable.

As usual, our starting point is a random experiment modeled by a probability space (Ω,F,P), so that Ω is the set of outcomes, F the collection of events, and PP the probability measure on the sample space (Ω,F). In this section, we assume that we fixed random variable Z defined on the probability space, taking values in IR.

Definition: For a ∈ IR and b∈(0,∞), let X=a + bZ . The two-parameter family of distributions associated with X is called the location-scale family associated with the given distribution of Z. Specifically, a is the location parameter and bb the scale parameter.

Thus a linear transformation, with positive slope, of the underlying random variable Z creates a location-scale family for the underlying distribution. In the special case that b=1, the one-parameter family is called the location family associated with the given distribution, and in the special case that a=0a=0, the one-parameter family is called the scale family associated with the given distribution. Scale transformations, as the name suggests, occur naturally when physical units are changed. For example, if a random variable represents the length of an object, then a change of units from meters to inches corresponds to a scale transformation.

Location transformations often occur when the zero reference point is changed, in measuring distance or time, for example. Location-scale transformations can also occur with a change of physical units. For example, if a random variable represents the temperature of an object, then a change of units from Fahrenheit to Celsius corresponds to a location-scale transformation.

1.3. PROBABILITY DENSITY FUNCTION

Random variables *X*: Let *S* be the sample space of an experiment. A real-valued function $X: S \to \mathbf{R}$ is called a random variable of the experiment if, for each interval $I \subseteq \mathbf{R}$, $\{s : X(s) \in I\}$ is an event.

The situation is similar for probability, except instead of calling it a mass density, we call it a probability density. The function that describes the probability density is called a probability density function, denoted by $f_X(x)$. Thinking carefully about the sand in the aquarium, one can see that probability density function can be expressed mathematically as

$$f_X(x) = \lim_{\Delta x \to 0} \frac{P[x \le X \le x + \Delta x]}{\Delta x} \qquad (1.1)$$

Probability density functions can take a variety of shapes. These are in fact two important properties of all probability density function satisfies two conditions:

1. $f_X(x) \ge 0$,
2. $\int_{-\infty}^{+\infty} f_X(x) = 1$.

Multiplying both sides by Δx, we obtain

$$\lim_{\Delta x \to 0} f_X(x)\Delta x = \lim_{\Delta x \to 0} P[x \le X \le x + \Delta x] \qquad (1.2)$$

or

$$f_X(x)dx = \lim_{\Delta x \to 0} P[x \le X \le x + \Delta x] \qquad (1.3)$$

In equation. (1.3), the "limit" operation enforces that Δx be small. However, we can lift this requirement, letting Δx be as large as we like, by summing the corresponding products on the left hand side through integration, i.e.,

$$\int_{x}^{x+\Delta x} f_X(x)dx = P[x \le X \le x + \Delta x] \qquad (1.4)$$

Letting $x = a$ and $x + \Delta x = b$ be any two real numbers, equation. (1.4) becomes

$$P[a \le X \le b] = \int_{a}^{b} f_X(x)dx \qquad (1.5)$$

Since the interval $a \le X \le b$ corresponds to an event, equation. (1.5) indicates that the probability of this event may be computed as the area under the curve of the probability density function in the interval. As a consequence of the fact that for continuous random variables $P[X = x] = 0$, the interval of integration may be closed or open, i.e.,

$$P[a \le X \le b] = P[a < X < b] = P[a \le X < b] = P[a < X \le b]$$

This is not the case if the random variable is of the mixed type, i.e., if it contains some discontinuities so that it is partially discrete. In this case, we use

$$P[a < X \leq b] = \int_a^b f_X(x)dx \qquad (1.6)$$

Where the lower inequality sign has been changed to < in order to exclude the influence that $P[X = a]$ might have if it is non-zero due to a discrete probability. Use of equation.

1.4. DISTRIBUTION FUNCTIONS

Definition: Suppose that X is a random variable with values in R. The (cumulative) distribution function of X is the function F:IR→[0,1] defined by

$$F(x)=P(X \leq x),\ x \in IR \qquad (1.7)$$

The distribution function is important because it makes sense for any type of random variable, regardless of whether the distribution is discrete, continuous, or even mixed, and because it completely determines the distribution of X; F(x) is the total probability mass to the left of (and including)

Figure 1.1: F(x) is the total probability to the left of (and including) x

Basic Properties: A few basic properties completely characterize distribution functions. Notationally, it will be helpful to abbreviate the limits of F from the left and right at x∈IR, Suppose that F is the distribution function of a real-valued random variable X.

1-F is increasing: if $x \leq y$ then $F(x) \leq F(y)$.

2-F(x+)=F(x) for x ∈ IR. Thus, F is continuous from the right.

3-F(x−)=P(X<x) for x∈ IR. Thus, F has limits from the left.

4-$F(-\infty)=0$.

5-$F(\infty)=1$.

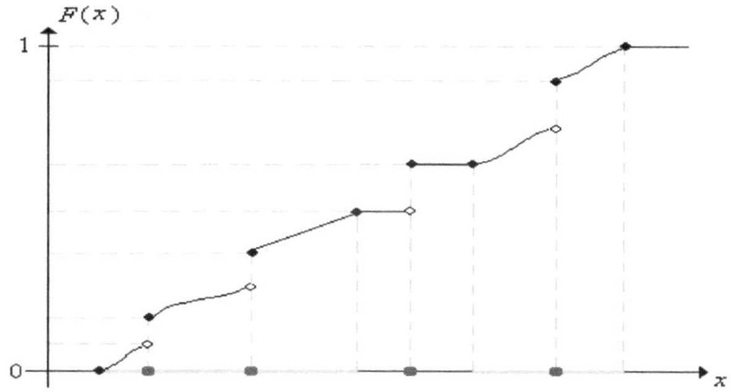

Figure 1.2: The graph of a distribution function

The following result shows how the distribution function can be used to compute the probability that X is in an interval. Recall that a probability distribution on R is completely determined by the probabilities of intervals; thus, the distribution function determines the distribution of X. Suppose again that F is the distribution function of a real-valued random variable X. If a, b $\in R$ with $a<b$ then

1-$P(X=a)=F(a)-F(a^-)$

2-$P(a< X \leq b)=F(b)-F(a)$

3-$P(a< X <b)=F(b^-)-F(a)$

4-$P(a\leq X\leq b)=F(b)-F(a^-)$

5-$P(a \leq X< b)=F(b^-)-F(a^-)$

Conversely, if a Function $F:R\rightarrow[0,1]$ satisfies the basic properties, then the formulas above define a probability distribution on R, with F as the distribution function. For more on this point, read the section on Existence and Uniqueness. If X has a continuous distribution, then the distribution function F is continuous.

1.5. RELATION TO DENSITY FUNCTIONS

There are simple relationships between the distribution function and the probability density function. Recall that if X takes value in S⊆IR and has probability density function f, we can extend f to all of R by the convention that f(x)=0 for x ∈ Sc. Suppose that X has discrete distribution on a countable subset S⊆IR. Let f denote the probability density function and F the distribution function. Thus, F is a step function with jumps at the points in S; the size of the jump at x is f(x).

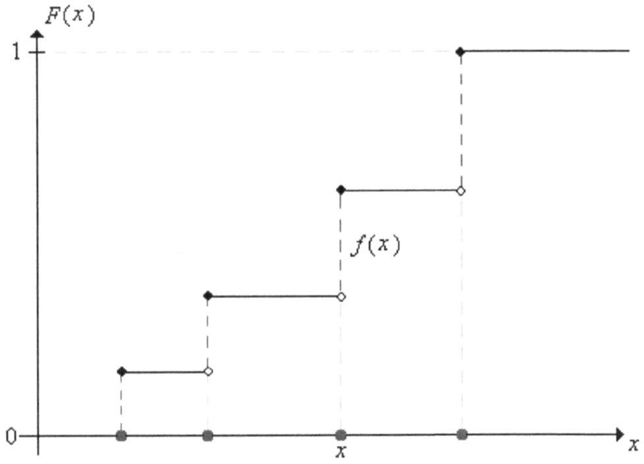

Figure 1.3: The distribution function of a a discrete distribution

There is an analogous result for a continuous distribution with a probability density function. Suppose that X has a continuous distribution on R with probability density function f and distribution function F.

1-F(x)=∫x−∞f(t)dt for x ∈IR.

2-f(x)=F′(x) if f is continuous at x.

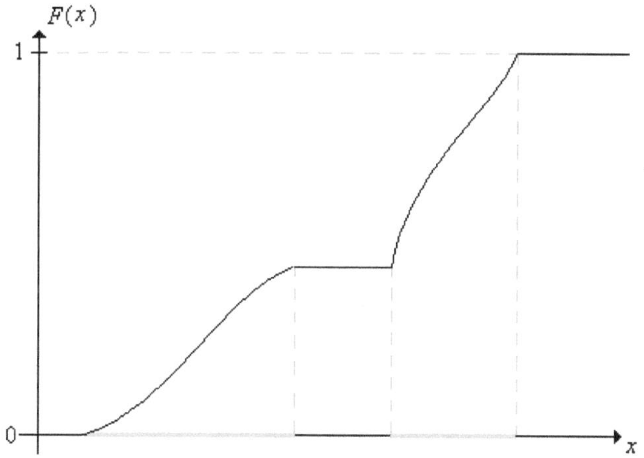

Figure 1.4: The distribution function of a continuous distribution

1.6. QUANTILE FUNCTION

The quantile function of a scalar random variable X is the inverse of its distribution function. The quantile function of a scalar random variable X is the inverse of its distribution function. Like the distribution function, the quantile function provides a complete description of the statistical properties of the random variable. For symmetrical distributions, the sample quantile function has a sigmoid shape, whereas for asymmetrical distributions the quantile function is convex or concave increasing.

Quantile-parameterized distributions are probability distributions that are directly parameterized by data. They were motivated by the need for easy-to-use continuous probability distributions flexible enough to represent a wide range of uncertainties, such as those commonly encountered in business, economics, engineering, and science.

Because quantile-parameterized distributions are directly parameterized by data, they have the practical advantage of avoiding the intermediate step of parameter estimation, a time-consuming process that typically requires non-linear iterative methods to estimate probability-distribution parameters from data. Some quantile-parameterized distributions have virtually unlimited shape flexibility and closed-form moments as well.

Specifically, we detail probabilistic features, such as the probability density, cumulative distribution and quantile functions, as well as lifetime indicators, such as the reliability or survival function and the failure or hazard rate. From the probability density and quantile functions, we are able to determine the mode and the median of the distribution. $F_X^{-1}(p)$ is the quantile function of the random variable X.

$$F_X^{-1}(p) = \inf\{x/F(x) \geq 1, 0 \leq p \leq 1\} \qquad (1.8)$$

The quantile-density function is also known as the sparsely function (Tukey (1965)). It is obtained as the first derivative of the quantile function with respect to p. It is defined as

$$f_X(u) = \frac{d\left(Q_X(u)\right)}{du} = Q'_X(u), 0 \leq u \leq 1\} \qquad (1.9)$$

As a result of the composite function $F_X(Q_X(u)) = u$ at arises from continuous random variables, the density quantile function is derived by taking derivatives on both sides of the function, such that

$$\frac{dF_X(Q_X(u))}{du} = \frac{du}{du} = 1$$
$$\Rightarrow f_X(Q_X(u))q_X(u) = 1$$
$$\Rightarrow f_X(Q_X(u)) = \frac{1}{q_X(u)}$$

It can be denoted as $f_X(Q_X(u))$ since it's derivative was obtained in terms of u and not X. The probability density curves from the Tukey lambda family of distributions exhibit various distributional shapes as a result of the kurtosis parameter. The quantile function, quantile density and density quantile functions of a real-valued random variable X, from the Tukey lambda distribution, are given as

$$Q_X(u) = \begin{cases} \mu + \frac{\sigma}{\lambda}\left[u^{\lambda} - (1-u)^{\lambda}\right]; \lambda \neq 0, 0 \prec u \prec 1, \\ \mu + \sigma \log\left(\frac{u}{1-u}\right); \lambda = 0, 0 \prec u \prec 1, \end{cases}$$

1.7.APPLICATIONS OF QUANTILE FUNCTION

Quantile functions are used in both statistical applications and Monte Carlo methods. The quantile function is one way of prescribing a probability distribution, and it is an alternative to the probability density function or probability mass function, the cumulative distribution function and the characteristic function.

The quantile function, Q, of a probability distribution is the inverse of its cumulative distribution function F. The derivative of the quantile function, namely the quantile density function, is yet another way of prescribing a probability distribution. It is the reciprocal of the the probability density function composed with the quantile function.

The original applications of Quantile parameterization were by decision analysts wishing to conveniently convert expert-assessed quintiles (e.g., 10th, 50th, and 90th quintiles) into smooth continuous probability distributions.

Quantile parameterization have also been used to fit output data from simulations in order to represent those outputs (both probability density distribution s and probability distribution function) as closed-form continuous distributions. Used in this way, they are typically more stable and smoother than histograms.

Similarly, since Quantile parameterization can impose fewer shape constraints than traditional distributions, they have been used to fit a wide range of empirical data in order to represent those data sets as continuous distributions (e.g., reflecting bimodality that may exist in the data in a straightforward manner . Quantile parameterization enables a closed-form representation of known distributions whose probability distribution function otherwise have no closed-form expression. Keelin et al. apply this to the sum of independent identically distributed lognormal distributions, where quintiles of the sum can be determined by a large number of simulations.

Nine such quintiles are used to parameterize a semi-bounded metalog distribution that runs through each of these nine quintiles exactly. Quantile parameterization have also been applied to assess the risks of asteroid impact, cyber security, biases in projections of oil-field production when compared to observed production after the fact, and future Canadian population projections based on combining the probabilistic views of multiple experts.

For statistical applications, users need to know key percentage points of a given distribution. For example, they require the median and 25% and 75% quartiles as in the example above or 5%, 95%, 2.5%, 97.5% levels for other applications such as assessing the statistical significance of an observation whose distribution is known; see the quantile entry.

Before the popularization of computers, it was not uncommon for books to have appendices with statistical tables sampling the quantile function. Statistical applications of quantile functions are discussed extensively by Gilchrist.

Functions implementing quantile methods can be found in common statistical software. A considerable number of R packages that provide such functions are available on the Comprehensive R Archive Network (CRAN). The base package stats contains basic functions to estimate sample quintiles or compute quintiles of common parametric distributions.

The quainter package is arguably a benchmark for distribution-free estimation of linear quantile regression models, as well as the base for other packages which make use of linear programming (LP) algorithms . Other contributions to the modeling of conditional quantile functions include packages for Bayesian regression, e.g. bayesQR and BSquare , and the lqmm package for random-effects regression.

The focus of this paper is on the R package Qtools, a collection of models and tools for quantile inference. These include commands for:

- quantile-based analysis of the location, scale and shape of a distribution;
    - transformation-based quantile regression;
    - goodness of fit and restricted quantile regression;
    - quantile regression for discrete data;

quantile-based multiple imputation. The emphasis will be put on the first two topics listed above as they represent the main contribution of the package, while a short description of the other topics is given for completeness.

Monte-Carlo simulations employ quantile functions to produce non-uniform random or pseudorandom numbers for use in diverse types of simulation calculations. A sample from a given distribution may be obtained in principle by applying its quantile function to a sample from a uniform distribution.

The demands of simulation methods, for example in modern computational finance, are focusing increasing attention on methods based on quantile functions, as they work well with multivariate techniques based on either copula or quasi-Monte-Carlo methods and Monte Carlo methods in finance.

1.8.EXERCISES

Exercise(1): The following exercise justifies the name $F^{-1}(p)$ is the minimum of the quintiles of order p. Let $p\in(0,1)$.

1-Prove that $F^{-1}(p)$ is a quantile of order p.

2- Prove that if x is a quantile of order p then $F^{-1}(p)\le x$.

Exercise(2): Let F be the function defined by

$$F^{-1}(u)=\begin{cases}\lambda^{-1}\left[u^{\lambda}-(1-u)^{\lambda}\right], & \lambda\neq 0\\ \\ (1-u)^{-1}\log(u), & \lambda=0\end{cases}$$

1-Sketch the graph of F and show that F is the distribution function for a discrete distribution.

2-Find the corresponding probability density function ff and sketch the graph.

3-Find $P(2\le X<3)$ where X has this distribution.

3-Find the quantile function and sketch the graph.

4-Find the five number summary and sketch the boxplot.

Exercise(3): Suppose that X has probability density function $f(x)=1/(b-a)$ for $x\in[a, b]$, where $a, b \in IR$ and $a<b$.

1-Find the distribution function and sketch the graph.

2-Find the quantile function and sketch the graph.

3-Compute the five-number summary.

4-Sketch the graph of the probability density function with the box plot on the horizontal axis.

Exercise(4): Suppose that X has probability density function $f(x)=re^{-rt}$ for $0\leq t<\infty$, where $r>0$ is a parameter.

1-Find the distribution function and sketch the graph.

2-Find the failure rate function and sketch the graph.

3-Find the quantile function and sketch the graph.

5-Compute the five-number summary.

6-Sketch the graph of the probability density function with the box plot on the horizontal axis.

Exercise(5): Let $f(x)=kx^k$ for $0<x<\infty$where $k>$is a parameter.

1. Sketch the graph of f in the cases $0<k<10$
2. Find the reliability function and sketch the graph.
3. Find the distribution function and sketch the graph.
4. Find the probability density function and sketch the graph.
5. Find the quantile function and sketch the graph.
6. Compute the five-number summary.

Exercise(6): Suppose that X has probability density function

$f(x)=12x^2(1-x)$ for $0\leq x\leq 10$.

1-Find the distribution function of X and sketch the graph.

2-Find $P(14\leq X\leq 12)$ and $P(6\leq X\leq 10)$.

3-Compute the five number summary and the inter quartile range. You will have to approximate the quintiles.

4-Sketch the graph of the density function with the box plot on the horizontal axis.

Exercise(7): Let $F(x)=e^x/(1+e^x)$ for $x \in IR$.

1-Show that F is a distribution function for a continuous distribution, and sketch the graph.

2-Compute $P(-1 \leq X \leq 1)$ and $P(0 \leq X \leq 10)$ where X is a random variable with distribution function F.

3-Find the quantile function and sketch the graph.

4-Compute the five-number summary and the inter quartile range.

5-Find the probability density function and sketch the graph with the box plot on the horizontal axis.

CHAPTER TWO:

TUKEY LAMBDA DISTRIBUTION

2.1.ORIGIN & HISTORY

The Tukey-λ distribution has interesting properties including (i) for some parameters values it has finite support, and for others infinite support, and (ii) it can mimic several other distributions such that parameter estimation for the Tukey distribution is a method for identifying an appropriate class of distribution to model a set of data.

The purpose of this work is to increase the methodological contributions on the mean difference and on the relationships of the mean difference with other variability indexes .

There are two general approaches to the study of income distributions, one motivated by the distribution of income among factors of production initiated by Ricardo (1817) and the other analyzing the distribution of total income of a given population receiving incomes making use of probability distribution functions.

The latter stream of thought introduced by Pareto (1895) has generated considerable interest among statisticians and economists to produce a large body of literature on new models based on empirical as well as theoretical justifications. The dynamic nature of income distributions always leaves scope for refinements to existing models and theories to make this a fertile area of research.

Several estimates of the parameters of a Tukey Lambda Distribution (TLD) have been devised, such as the method of moments (Pearson, 1894). The lambda distribution is a continuous probability distribution defined in terms of its quantile function.

It is a shape-conformable distribution used to identify an appropriate common distribution family to fit a collection of data to. The studies on the mean difference, introduced by Corrado Gini in 1912 as a measure of the variability of the characters according to the aspect of inequality, have aroused the interest of many scholars over years and also recently.

The importance of mean difference is also due to the fact that the sample mean difference is a correct estimate of that of the population distribution model and, therefore, functional for inferential purposes.

The theoretical contributions on the mean difference concern the main continuous distribution models (normal, rectangular, exponential, ...), however, for other distribution models, such as Tukey's, no contributions are known in literature.

The Lambda distribution originally derived as the one-parameter lambda distribution and proposed by Tukey. Tukey's lambda distribution was generalized, for the purpose of generating random variates for Monte Carlo simulation studies.

The extension of Tukey's lambda distribution [Hastings et al., 1947; Tukey, 1960, 1962]. Tukey's lambda distribution is symmetric and is defined by its quantile function. It was introduced to efficiently redistribute uniform random variates into approximations of other distributions [Tukey, 1960; Van Dyke, 1961].

Soon after the introduction of Tukey's lambda distribution, Hogben [1963], Shapiro et al. [1968], Shapiro and Wilk [1965], and Filliben [1975] introduced the non-symmetric case in sampling studies. Over the years, Tukey's lambda distribution became a versatile distribution, with location, scale, and shape parameters, that could accommodate a large range of distribution forms.

Tukey lambda distribution is usually used to choose a distribution model to fit data and its direct use is less usual. In general, its characteristic is that neither its density function f(x) nor its cumulative function F(x) is known, but only the inverse of this latter $F^{-1}(x)$, that is the quantile function Q(p).

A complete Tukey distribution shape includes three parameters: one of position, one of scale and one of shape. In order to calculate the mean difference and the mean deviation, it is better to refer to a reduced distribution in which the position parameter is set to zero and the scale to one. Formulas of mean difference and mean deviation of complete distribution are equal to the ones of reduced distribution multiplied by the scale parameter value.

The Tukey lambda distributions (Hastings *et al.* 1947; Tukey, 1962; Shapiro & Wilk, 1965) have been found useful in a number of univariate sampling problems. Tukey lambda distribution is defined by the quantile function;

$$x = F^{-1}(u) = \frac{\left[u^{\lambda} + (1-u)^{\lambda}\right]}{\lambda}; \lambda \neq 0, 0 \prec u \prec 1, \qquad (2.1)$$

Said function is not always analytically invertible and, therefore, allows to obtain cumulative function and density function only for some values of λ. It is necessary to use numerical inversion of Q(p) to get a cumulative function for other λ values.

Regarding Tukey distribution, some characteristic values as function of λ are known: average, mode, median, standard deviation, asymmetry index, disnormality excess index, entropy, characteristic function. Expressions of mean difference and mean deviation are unknown. Several samples of the PDFs of the Tukey lambda distributions.

A modest beginning to the latter law of thought was made by Tarsitano (2004) when he used a general version of the Tukey Lambda family of quantile distributions to model income data. Different schools of thought prevail in forming the attitude towards analysis of statistical data, broadly classified as confirmatory and exploratory analyses.

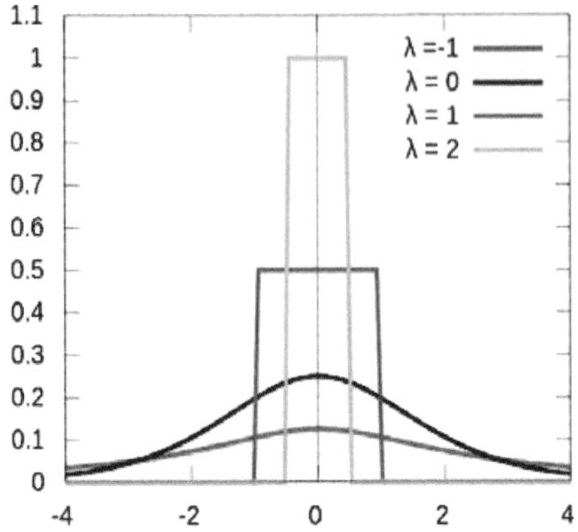

Figure 2.1: The graph of a Tukey Lambda distribution function

Several authors have considered the problem of deriving 8ayes estimates for the parameters of the Pareto type I distribution. [Malik (1970). Zellner (1971). Lwin (1972). Rao Tummala (1977) and Sinha and Howlader (1980)]. Silcock (1954) and Harris (1968) proposed maximum likelihood estimates, Harris (1968) (when,u =o) and Arnold and Laguna (1977) used the method of moments, while Moore and Harter (1967, 1969), Kulldorff and Vannman (1973) and Vannman (1976) suggested estimates based on order statistics for the parameters of Pareto(lI) distribution.

The properties of the above distributions, interpretations for the parameters and all the above estimation techniques are discussed in detail in Arnold (1983). Classical regression type estimators and several recent developments. notably in connection with UMVU estimation are discussed in Kleiber and Kotz (2003). An alternative way of describing a continuous distribution is to use the quantile functions.

In particular, we model the marginal distributions with the Generalized Tukey lambda (GTL) distribution. First proposed by Pregibon (1980) and Freimer et al. (1988), the GTL distribution is very flexible, allowing varying degrees of asymmetry and tail thickness. For example, it nests the logistic and uniform distributions and approximately nests familiar distributions such as the normal, Student's t, Gumbel, and χ 2 distributions. Because of this versatility, the GTL is a good candidate distribution for modeling the unobservable in the selection and outcome equations.

These GTLs are then inserted as marginal distributions into a copula in order to construct a highly versatile bivariate distribution that replaces the bivariate normality distribution that underlies the approach of Heckman (1979) and Lee (1978).

Probability density function does not have closed form but can be found numerically:

```
In[1]:= Plot[Table[PDF[TukeyLambdaDistribution[λ], x], {λ, {-1, 0.5, 1}}] // Evaluate,
          {x, -3, 3}, Filling → Axis]
```

Out[1]=

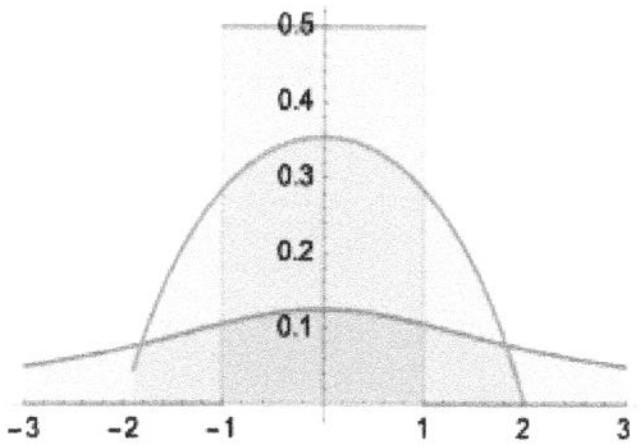

Cumulative distribution function does not have closed form but can be found numerically:

```
In[1]:= Plot[Table[CDF[TukeyLambdaDistribution[λ], x], {λ, {-1, 0.5, 1}}] // Evaluate,
          {x, -3, 3}, Exclusions → None, Filling → Axis]
```

Out[1]=

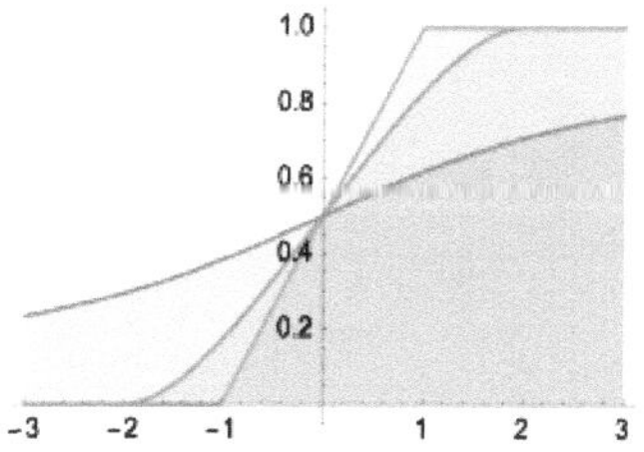

Figure 2.2: The graph of a Tukey Lambda distribution function

The flexibility of this GTL-copula distribution effectively frees the sample selection model from any particular distributional assumption. Moreover, this flexibility is achieved with just a few additional parameters, which is both parsimonious and time-efficient relative to semi- or non-parametric approaches.

Cumulative functions of Tukey lambda distribution for such values are listed in graph. Therefore, allows to obtain cumulative function and density function only for some values of λ. In this work, the formulas of mean difference and mean deviation of Tukey Lambda distribution have been obtained.

Quantile function for symmetric case:

```
In[1]:= Plot[Table[Quantile[TukeyLambdaDistribution[λ], q], {λ, {-2, -1, 0.5}}] // Evaluate,
          {q, 0, 1}, Filling -> Axis]
```

Out[1]=

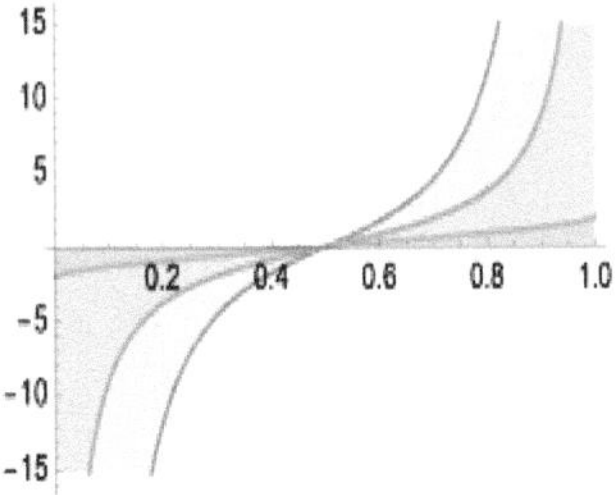


```
In[1]:= Quantile[TukeyLambdaDistribution[λ], q]
```

Out[1]=

$$\begin{cases} \begin{cases} \frac{-(1-q)^{\lambda}+q^{\lambda}}{\lambda} & \lambda \neq 0 \\ -\text{Log}[1-q]+\text{Log}[q] & \text{True} \end{cases} & 0<q<1 \\ \begin{cases} -\infty & \lambda \leq 0 \\ -\frac{1}{\lambda} & \text{True} \end{cases} & q \leq 0 \\ \begin{cases} \infty & \lambda \leq 0 \\ \frac{1}{\lambda} & \text{True} \end{cases} & \text{True} \end{cases} \quad \text{if } 0 \leq q \leq 1$$

Figure 2.3: The graph of a Tukey Lambda distribution function

It is an original contribution aimed at increasing the knowledge about this distribution model. These results allowed us to investigate the relationships among the three main variability indexes, standard deviation, mean deviation and mean difference, regarding Tukey lambda model.

The Tukey-λ is, however, symmetric. Here we define a new class of represents Tukey's lambda distribution with shape parameter λ distribution with similar properties to the Tukey-λ distribution. As with the Tukey-λ distribution, our distribution is defined in terms of its quantile function, which in this case is given by the polylogarithm function.

The Tukey lambda distribution is a continuous probability distribution defined in terms of its quantile function. It is typically used to identify an appropriate distribution (see the comments below) and not used in statistical models directly.

The Tukey lambda distribution has a shape parameter . As with other probability distributions, the Tukey lambda distribution can be transformed with a location parameter,, and a scale parameter, . Since the general form of probability functions can be expressed in terms of the standard distribution, the subsequent formula is given for the standard form of the function.

In[1]:=
```
Plot[Table[Quantile[TukeyLambdaDistribution[1.5, 1, σ], q], {σ, {0.5, 1, 2}}] // Evaluate,
  {q, 0, 1}, Filling -> Axis]
```

Out[1]=
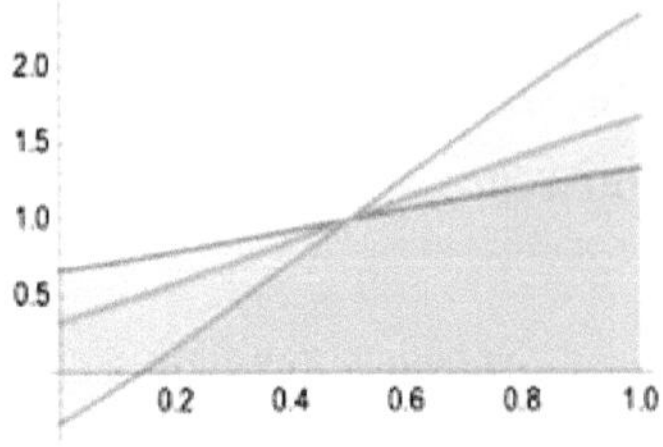

In[2]:=
```
Plot[Table[Quantile[TukeyLambdaDistribution[1.5, μ, 1.5], q], {μ, {-0.5, 1, 2}}] // Evaluate,
  {q, 0, 1}, Filling -> Axis]
```

Out[2]=
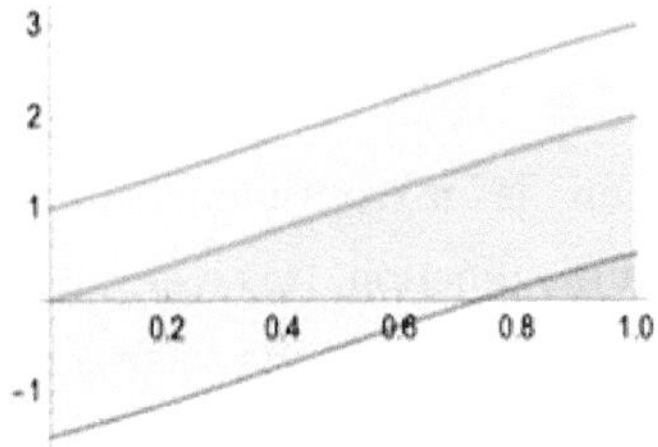

In[3]:=
```
Quantile[TukeyLambdaDistribution[λ, μ, σ], q]
```

Figure 2.4: The graph of a Tukey Lambda distribution function

The Qut[3] of Tukey lambda distribution as a function of λ parameter is;

Out[3]=

$$\begin{cases} \begin{cases}\mu+\frac{\left(-(1-q)^{\lambda}+q^{\lambda}\right)\sigma}{\lambda} & \lambda\neq 0\\ \mu+\sigma(-\text{Log}[1-q]+\text{Log}[q]) & \text{True}\end{cases} & 0<q<1\\ \begin{cases}-\infty & \lambda\leq 0\\ \mu-\frac{\sigma}{\lambda} & \text{True}\end{cases} & q==0\\ \begin{cases}\infty & \lambda\leq 0\\ \mu+\frac{\sigma}{\lambda} & \text{True}\end{cases} & \text{True}\end{cases} \quad \text{if } 0\leq q\leq 1$$

We show the support of the distribution to be the Riemann zeta function (when finite), and we provide a closed form for the expectation, provide simple means to calculate the CDF and PDF, and show that it has relationships to the uniform, exponential, inverse beta and extreme-value distributions, represents Tukey's lambda distribution with location parameter μ and scale parameter σ.

We will discuss in detail the family of Tukey transform models, then introduce our estimation method, and finally present an empirical application using a real world dataset.

2.2. GENERAL FAMILIES OF QUANTILE TRANSFORM DISTRIBUTIONS

In next time The Tukey-λ distribution has represents the generalized Tukey's lambda distribution with location parameter μ, scale parameters σ_1 and σ_2, and shape parameters λ_1 and λ_2.

Median of generalized Tukey lambda distribution:

```
In[1]:= Median[TukeyLambdaDistribution[{λ1, λ2}, μ, {σ1, σ2}]]
```

Out[1]=

$$\begin{cases} \mu + \frac{(-1+2^{-\lambda 1})\sigma 1}{\lambda 1} - \frac{(-1+2^{-\lambda 2})\sigma 2}{\lambda 2} & \lambda 1 \neq 0 \,\&\&\, \lambda 2 \neq 0 \\ \mu - \frac{(-1+2^{-\lambda 2})\sigma 2}{\lambda 2} - \sigma 1 \, \mathrm{Log}[2] & \lambda 1 == 0 \,\&\&\, \lambda 2 \neq 0 \\ \mu + \frac{(-1+2^{-\lambda 1})\sigma 1}{\lambda 1} + \sigma 2 \, \mathrm{Log}[2] & \lambda 1 \neq 0 \,\&\&\, \lambda 2 == 0 \\ \mu - \sigma 1 \, \mathrm{Log}[2] + \sigma 2 \, \mathrm{Log}[2] & \mathrm{True} \end{cases}$$

Probability density function with location and scale:

```
In[1]:= Plot[Table[PDF[TukeyLambdaDistribution[0.4, -1, σ], x], {σ, {0.5, 1, 2}}] // Evaluate,
         {x, -4, 2}, Filling → Axis]
```

Out[1]=

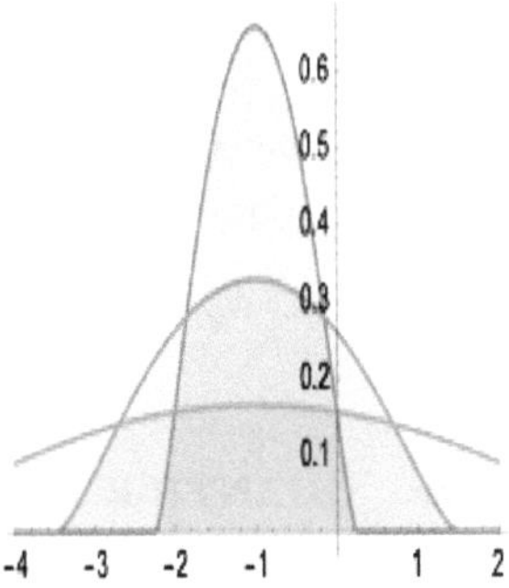


```
In[2]:= Plot[Table[PDF[TukeyLambdaDistribution[λ, -1, 0.5], x], {λ, {-0.5, 0.2, 0.7}}] // Evaluate,
         {x, -4, 2}, Filling → Axis]
```

Out[2]=

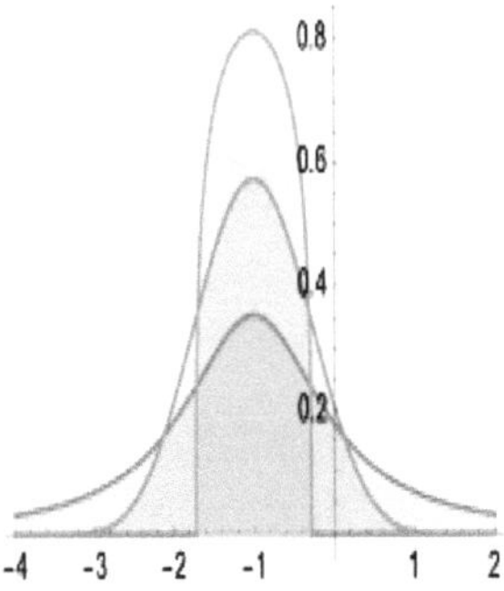

Figure 2.6: The graph of a Tukey Lambda distribution function

Quantile function for generalized Tukey's lambda distribution:

```
In[1]:= Plot[Table[Quantile[TukeyLambdaDistribution[{5, 2}, μ, {1, 2}], q], {μ, {-3, -1, 0.5}}] // Evaluate,
         {q, 0, 1}, Filling -> Axis]
```

Out[1]=

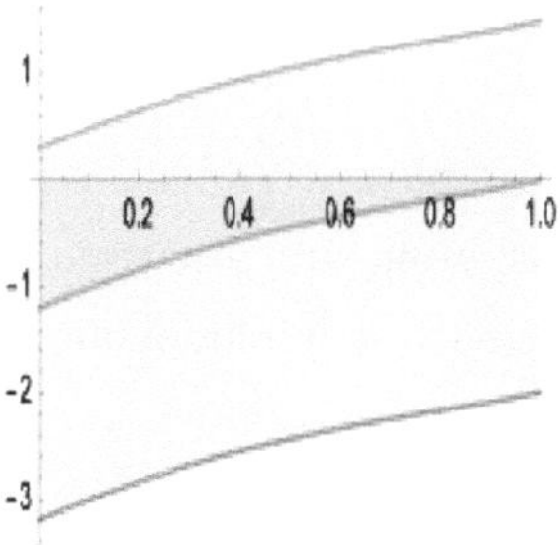

Figure 2.7: The graph of a Tukey Lambda distribution function

It is well known that distributions of financial returns can exhibit heavy-tails, skewness, or both. The Tukey Lambda Distribution has two parameterizations, one due to Ramberg & Schmeiser (RS), and the other to Freimer, Mudholkar, Kollia and Lin (FMKL).

What's remarkable about this distribution, which is defined by just four parameters (canonically referred to as lambda 1 through 4), is that it can approximate the Normal, Uniform, Beta, Students-t, Exponential, Weibull, Gamma, Lognormal, Pareto, F, Chi-square, Logistic, Double Exponential and Extreme Value distributions, depending on the values that its quartet of lambdas take on.

Tukey's lambda is a symmetric distribution that is defined by its quantile function. It was introduced to efficiently generate random variates that approximate other distributions. Soon after the introduction of Tukey's lambda distribution, Hogben, Shapiro et al., Shapiro and Wilk , and Filliben used the non-symmetric case in sampling studies.

we introduce a new evolutionary optimization algorithm based on Tukey's symmetric lambda distribution. Tukey distribution is defined by 3 parameters, the shape parameter, the scale parameter, and the location parameter or average value. Various other distributions can be approximated by changing the shape parameter, and as a result can encompass a large class of probability distributions.

In addition, Because of these attributes, an Evolutionary Programming (EP) algorithm with Tukey mutation operator may perform well in a large class of optimization problems. Various schemes in implementation of EP with Tukey distribution are discussed, and the resulting algorithms are applied to selected test functions and antenna design problems.

The Tukey lambda family of distributions together with its extensions have played an important role in statistical practice. In this paper a continuously defined two-parameter generalization of this family, which holds promise of a variety of additional applications, is variously studied. The coefficients of skewness and kurtosis and the density shapes of its members are examined and the family is related to the classical Pearsonian system of distributions.

Over the years, it became a versatile distribution with location, scale, and shape parameters that can accommodate a large range of distribution shapes. It was used to analyze data and was no longer restricted to approximating other distributions.

Figures illustrates the four different shapes of the TLD: unimodal, U-shape, Monotone and S-shape. In applications, the TLD has been used in various fields, such as in option pricing as a fast generator of financial prices or in fitting income data .

It has also been used in meteorology , in studying the fatigue lifetime prediction of materials, in simulations of queue systems, in corrosion studies, and in independent component analysis. Finally, it has been used in statistical process control .

We begin by discussing the general family of Tukey distributions. Basically, Tukey suggested several nonlinear transformations of a standard Gaussian random variable, denoted below throughout by W~Normal(0,1).

The g-and-h transformations involve a skewness transformation of type g and a kurtosis transformation of type h. If one replaces the kurtosis transformation of the type h with the type k, one obtains the g-and-k family of distributions discussed by more authors. If the type h transformation is replaced by the type j transformation, one obtains the g-and-j transformations .

Cumulative distribution function with location and scale:

```
In[1]:= Plot[Table[CDF[TukeyLambdaDistribution[0.4, -1, σ], x], {σ, {0.5, 1, 2}}] // Evaluate,
          {x, -4, 2}, Filling → Axis]
```

Out[1]=

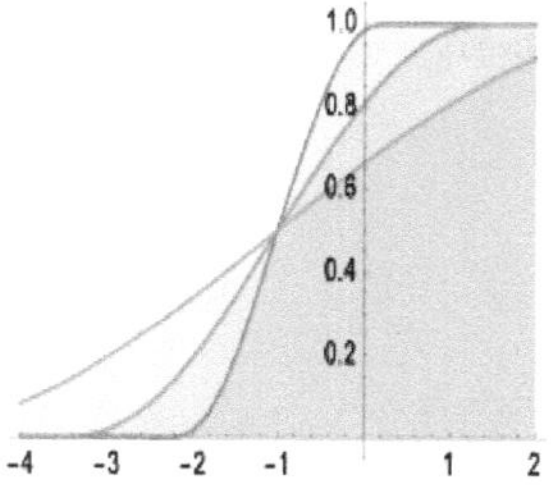


```
In[2]:= Plot[Table[CDF[TukeyLambdaDistribution[λ, -1, 0.5], x], {λ, {-0.5, 0.2, 0.7}}] // Evaluate,
          {x, -4, 2}, Filling → Axis, Exclusions → None]
```

Out[2]=

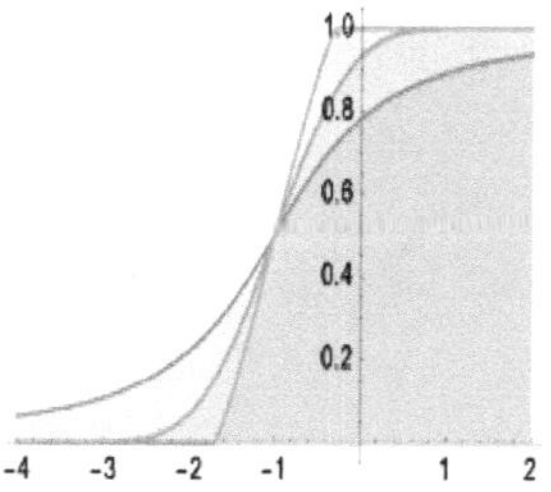

Figure 2.8: The graph of a Tukey Lambda distribution function

This trait is often taken as stylized fact, and can be modeled by α-stable distributions. The drawback to using these distributions is that they do not all have closed form for their densities and distribution functions.

It gained use as a data analysis tool and was no longer restricted to approximating other distributions. Figure 1 illustrates the four qualitatively different shapes of the TLD: unimodal, U-shaped, monotone and S-shaped. The TLD has been applied to numerous fields, including option pricing as a fast generator of financial prices, and in fitting income data . It has also been used in meteorology , in studying the fatigue lifetime prediction of materials, in simulations of queue systems, in corrosion studies , in independent component analysis, and in statistical process control .

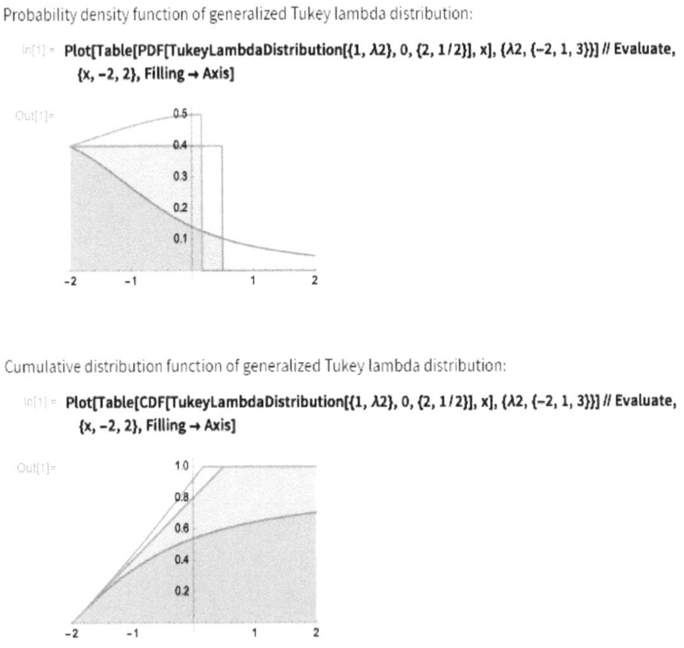

Figure 2.9: The graph of a Tukey Lambda distribution function

2.3.VARIABILITY INDEXES OF TUKEY LAMBDA DISTRIBUTION

Regarding Tukey distribution, some characteristic values as function of λ are known: average, mode, median, standard deviation, asymmetry index, disnormality excess index, entropy, characteristic function. Expressions of mean difference and mean deviation are unknown. The variance of Tukey lambda distribution as a function of λ parameter is;

$$\sigma^2 = \frac{2}{\lambda^2}\left[\frac{1}{1+2\lambda} - \frac{\Gamma(\lambda+1)^2}{\Gamma(2\lambda+2)}\right]; \lambda \succ -\frac{1}{2}, \qquad (2.2)$$

By using the cumulative functions derived by the inversion of quantile functions of Tukey lambda distribution, mean difference and mean deviation values are obtained.

Tukey sub-families of loss models and in particular different ways that people have sought to evaluate and present the distribution and density functions for the popular sub-families such as the g-and-h, generalized g-and-h and g-and-k families. In general, it will be informative for this section to remind the reader of the following basic property.

If X is a continuous random variable distributed according to distribution X~Fx(x), which is monotonically increasing on support Supp{Fx(x)}={x:0<Fx(x)<1}, then, in this general case, one can show that the quantile function $Qx(\alpha)=F^{-1}x(\alpha)$ for $\alpha \in [0,1]$ determines the relationship between the random variable X and any other continuous random variable with monotonically increasing distribution, say W~Fw(w), through the relationship given as follows

$$Q_X(\alpha) = F_X^{-1}(F_W(Q_W(\alpha))) \qquad (2.3)$$

Furthermore, the following relationship between the quantile function of a random variable X and its density can be obtained by using the identity for differentiation of an inverse function given by;

$$\frac{d}{dx}g^{-1}(x)=\left[\frac{d}{dx}g\left(g^{-1}(x)\right)\right]^{-1} \qquad (2.4)$$

This result, when applied to the quantile function of the random variable X, produces the following relationship;

$$\frac{dQ_X(\alpha)}{d\alpha}=\frac{1}{f_X(Q_X(\alpha))} \qquad (2.5)$$

where Qx(α) is the quantile function for random variable X at quantile level α, and fx(·) represents the density for random variable X. One can then also apply this to the relationship to obtain;

$$\frac{dQ_X(\alpha)}{d\alpha}=\frac{d}{du}Q_X(u)\frac{d}{d\alpha}=\frac{f_X(Q_W(\alpha))}{f_X\left(F_X^{-1}\left(F_W\left(Q_W(\alpha)\right)\right)\right)} \qquad (2.6)$$

The density functions for different members of the Tukey family of quantile specified loss models. In the following results we basically apply the same methodology to obtain the density and distribution for each of the different Tukey classes of loss model.

We begin with the density of the super class of transformations presented previously according to the quantile transformation. The symmetrical Tukey lambda distributions are obtained by simple transformations of a uniformly distributed variable.

Systems of multivariate distributions can be formed by applying these transformations to sets of variables having a joint Dirichlet distribution. Since no more than one of such a set of variables can have a uniform distribution, though all have beta distributions, we are led to study distributions of transforms of variables having standard beta distributions. These distributions are termed extended Tukey lambda distributions. Properties of these distributions are studied.

2.4. ESTIMATION OF THE TUKEY LAMBDA DISTRIBUTION PARAMETER

Estimation of the lambda distribution parameter is not straightforward, and available estimators that perform best have large computation times. The considered methods are Method of Probability Density Quintiles(pdQ), Method of TL-moments (TL), Percentile Matching Method (PM,) Numerical Maximum Likelihood (ML), Maximum Product of Spacings (MPS), Titterington's Method (TM), Starship Method (SM), and Method of Distributional Least Absolutes (DLA).

All the methods display coverage probabilities close to the nominal value for the location parameter. The TM method has the narrower width, with the pdQ method the next narrowest. We found that larger sample sizes are required to obtain reliable intervals for the difference in shape parameters.

The TM method is not considered here as BCa bootstrap confidence intervals need a higher number of re-samples and when the sample size increases to 1000, and it takes roughly 33.98 minutes to obtain a single confidence interval for TM. As can be seen, both TL-moments and pdQ method provide the best fit to the data.

Further, this is elaborated in Figure 4 quantile plots where, unlike the other methods, estimated quintiles from the pdQ method and TL-moments method are clearly aligned with the sample quintiles, we provide the running times for parameter estimation by each method. The PM and pdQ methods are clearly the fastest.

The pdQ is about five times faster than the TL-moments method and 91 times faster than SM for this particular data set. This further suggests the suitability of the pdQ method with large sample sizes.

The pdQ and PM methods, on the other hand, take only 47.65 seconds and 32.02 seconds subsequently to provide a single confidence interval for n = 1000. Compared to the PM method, the pdQ method shows better coverage in the vicinity of the nominal level for skewness with a narrower width. So it can be concluded that the pdQ is favorable compared to these other two methods.

2.5.APPLICATIONS OF TUKEY LAMBDA DISTRIBUTION

The Tukey-Lambda distribution is typically used to identify an appropriate distribution (see the comments below) and not used in statistical models directly. For this reason, we omit the formulas, and plots for the hazard, cumulative hazard, survival, and inverse survival functions. We also omit the common statistics and parameter estimation sections.

The most common use of this distribution is to generate a Tukey-Lambda PPCC plot of a data set. Based on the ppcc plot, an appropriate model for the data is suggested. For example, if the maximum correlation occurs for a value of λ at or near 0.14, then the data can be modeled with a normal distribution. Values of λ less than this imply a heavy-tailed distribution (with -1 approximating a Cauchy).

That is, as the optimal value of λ goes from 0.14 to -1, increasingly heavy tails are implied. Similarly, as the optimal value of λ becomes greater than 0.14, shorter tails are implied.

As the Tukey-Lambda distribution is a symmetric distribution, the use of the Tukey-Lambda PPCC plot to determine a reasonable distribution to model the data only applies to symmetric distributions. A histogram of the data should provide evidence as to whether the data can be reasonably modeled with a symmetric distribution.

The most common use of this distribution is to generate a Tukey lambda PPCC plot of a data set. Based on the PPCC plot, an appropriate model for the data is suggested. For example, if the best-fit of the curve to the data occurs for a value of λ at or near 0.14, then the data could be well-modeled with a normal distribution.

In this paper we have introduced a simple two-step estimation method for the TLD parameters. The first step uses the empirical probability density quantile function to find estimates of the shape parameters. The second step can then simply obtain estimates for the location and inverse scale. How simulations show that the method performs very well, often beating existing methods. An additional advantage of the approach is the very small computation time compared to the best of the methods.

This small computation time makes bootstrapping possible which is important given the lack of interval estimators available for the TLD parameters. For the more time consuming estimation methods, bootstrap intervals can take hours, even days to compute, as our examples show. While we focused on the TLD distribution in this paper, the method can be used for other distributions too. An obvious extension would be to consider other TLD forms, like the generalized Beta distribution, since estimation for such distributions is not straightforward.

Values of λ less than 0.14 suggests a heavier-tailed distribution; a milepost at $\lambda = 0$ (logistic) would indicate quite fat tails, with the extreme limit at $\lambda = -1$, approximating Cauchy.

That is, as the best-fit value of λ varies from 0.14 towards -1, a bell-shaped PDF with increasingly heavy tails is suggested. Similarly, for an optimal value of λ becomes greater than 0.14 suggests a distribution with exceptionally thin tails (based on the point of view that the normal distribution itself is thin-tailed to begin with). Except for values of λ very close to 0, all the suggested PDF functions have finite support, between $-1/|\lambda|$ and $+1/|\lambda|$.

Since the Tukey lambda distribution is a symmetric distribution, the use of the Tukey lambda PPCC plot to determine a reasonable distribution to model the data only applies to symmetric distributions. A histogram of the data should provide evidence as to whether the data can be reasonably modeled with a symmetric distribution.

The limitations of classical hypothesis testing methodology with regard to the discrimination between Gaussian and non-Gaussian data were firstly highlighted through an original statistical example.

As far as SPC is concerned, an alternative methodology is proposed to leave aside the nature of the unknown theoretical underlying distribution characterizing the process and hence to avoid questionable inference hypothesis.

As the process is stopped when an abnormal pattern is detected in order to find an assignable cause, this means that the steel production would have been stopped unnecessarily five times in a week if the chloride concentration had been modeled using a lognormal law.

Knowing the industrial cost of a production interruption, this example illustrates the economic nonsense of using the SPC methodology without an appropriate statistical modeling of the process under study.

In the previous example, the eight Western Electric rules were tested together, leading to an overall probability of a false alarm (Type I error) close to a = 2.23% (see Appendix for details), which is of course an intolerably high false alarm rate for most applications. In practical SPC implementations, a choice of some of these different rules is often made, and exact calculation or simulation of the false alarm rates for usual combinations can be performed.

We now consider a much smaller data set known as the Indiana Twin Study data, which consists of the birth weights of a set of 123 twins. The data originally comes from the PhD thesis of Dr Cynthia Moore, Department of Medical and Molecular Genetics.

According to the p-values, all the methods suggest a better fit to the data, but in particular, the SM method, TL-moments method and pdQ method are the best fits, as they display smaller KS statistics compared to the other methods.

A rules-selection feature is included in the LambdaSPC software in order to try different possible combinations. Nevertheless, if one wants, for a given combination of rules, to limit the overall false alarm rate to a chosen value, it is necessary to adjust these rules. Such an adjustment for multiple testing can be performed by means of a Bonferroni-type procedure as advocated in the multivariate case by Alt et al. or using some other method .

The use of the TLD modeling has no incidence on these procedures, except that it eases the modification of the SPC limits thanks to the percentile-based definition of the TLD.

Such a methodology is based on the use of the TLD; a very flexible family of distributions that fit to a broad class of distributions including the most usual ones (uniform, lognormal, Gaussian, Weibull,...) while keeping a very accurate precision on the predicted control limits used in SPC methodology.

Besides, it was shown that the minimal size of the initial dataset required to obtain an acceptable accuracy on the predicted SPC limits is only of approximately a hundred measurements when considering TLD-based modeling tested in a particularly unfavorable case.

Finally, the relevance of using the TLD in SPC was illustrated treating experimental data originating from an industrial production line using the well-known Western Electric rules.

Despite the mathematical complexity of determining the TLD parameters, this statistical family is made easy to use in SPC methodology thanks to the Lambda Finder and Lambda SPC software we especially developed to perform such a task. Further information on these software and a demo-version can be sent by mail by the authors upon request.

The Tukey lambda family of distributions together with its extensions have played an important role in statistical practice. In this paper a continuously defined two-parameter generalization of this family, which holds promise of a variety of additional applications, is variously studied.

The coefficients of skewness and kurtosis and the density shapes of its members are examined and the family is related to the classical Pearsonian system of distributions.

2.6.EXERCISES

Exercise(1): Let X be the Tukey-Lambda continuous random variable. It is inherited from the of generic methods as an instance of the random variable continuous class. It completes the methods with details specific for this particular distribution. defined by:

$$F^{-1}(p)=\lambda^{-1}\left(p^{\lambda}-(1-p)^{\lambda}\right\},\lambda\neq 0 \ ,$$

$F^{-1}(p)$ is the minimum of the quintiles of order p. Let p∈(0,1).

1-Prove that $F^{-1}(p)$ is a quantile of order p.

2- Prove that if x is a quantile of order p then $F^{-1}(p)\leq x$.

Exercise(2): Let F be the closed form of Tukey-Lambda distribution function defined by

$$F^{-1}(u)=\begin{cases}\lambda^{-1}\left[u^{\lambda}-(1-u)^{\lambda}\right], & \lambda\neq 0\\ \\ (1-u)^{-1}\log(u), & \lambda=0\end{cases}$$

1-Sketch the graph of F and show that F is the distribution function for a discrete distribution.

2-Find the Maximum Entropy Estimation of density Tukey-Lambda distribution function

3-Find the Quantile function where X has this distribution.

3-prove that $\begin{cases}f(x)\geq 0\\ \int_{Q(0)}^{Q(1)}f(x)dx=1\end{cases}$

Exercise(3): Let X be distributed as standard uniform U(0,1), let the transformation

$$Y=\begin{cases}\lambda^{-1}\left(aX^{\lambda}-(1-X)^{\lambda}\right\},\lambda\neq 0\\ \log\left(\dfrac{X^{a}}{1-X}\right),a\succ 0\end{cases}$$

1-Find the distribution function of Y and sketch the graph.

2-Find the quantile function and sketch the graph.

3-Compute the five-number summary.

4-Sketch the graph of the probability density function with the box plot on the horizontal axis.

Exercise(4): prove for the standard form of the Tukey lambda distribution, the quantile function, Q(p) (i.e. the inverse function to the cumulative distribution function) and the quantile density function d Q/dp are

$$Q_X(p,\lambda)=\begin{cases}\frac{1}{\lambda}\left[p^{\lambda}-(1-p)^{\lambda}\right], if\ \lambda\neq 0,\\ \log\left(\frac{p}{1-p}\right), if\ \lambda=0,\end{cases}$$

$$q_X(p,\lambda)=Q_X(p,\lambda)/dp=\left[p^{(\lambda-1)}-(1-p)^{(\lambda-1)}\right].$$

Exercise(5): For the Tukey lambda distribution is symmetric around zero, therefore the expected value of this distribution is equal to zero. The variance exists for $\lambda > -1/2$ and is given by the formula (except when $\lambda = 0$).Prove then:

1- $\mu_n = E[X^n]=\lambda^{-n}\sum_{k=0}^{n}(-1)^k\binom{n}{k}B(\lambda k+1,\lambda(n-k)+1).$

2- $Var[X]=\frac{2}{\lambda^2}\left(\frac{1}{1+2\lambda}-\frac{\Gamma(\lambda+1)^2}{\Gamma(2\lambda+2)}\right).$

Exercise(6): Differently from the central moments of the Tukey lambda distribution. Prove that the L-moments can be expressed in a closed form. The L-moment of order r>1 is given by

$$L_n=E[X^n]=\lambda^{-1}\left[1+(-1)^r\sum_{k=0}^{r-1}(-1)^{r-1-k}\binom{r-1}{k}\binom{r+k-1}{k}\left(\frac{1}{k+1+\lambda}\right)\right.$$

The first six L-moments can be presented as follows:

$$\begin{cases} L_1 = 0 \\ L_2 = \dfrac{2}{\lambda}\left[-\dfrac{1}{1+\lambda} + \dfrac{2}{2+\lambda}\right] \\ L_3 = 0 \\ L_4 = \dfrac{2}{\lambda}\left[-\dfrac{1}{1+\lambda} + \dfrac{12}{2+\lambda} - \dfrac{30}{3+\lambda} + \dfrac{30}{4+\lambda}\right] \\ L_5 = 0 \\ L_6 = \dfrac{2}{\lambda}\left[-\dfrac{1}{1+\lambda} + \dfrac{30}{2+\lambda} - \dfrac{210}{3+\lambda} + \dfrac{560}{4+\lambda} - \dfrac{630}{5+\lambda} + \dfrac{252}{6+\lambda}\right]. \end{cases}$$

CHAPTER THREE:

THE GENERALIZED TUKEY LAMBDA DISTRIBUTION

3.1.ORIGIN & HISTORY

Historically, The original parameterization, referred to as the RS generalized Tukey-Lambda distribution the The First Generalized Tukey Lambda Distribution is done by Tukey, and the Tukey's _ family of distribution is defined by the quantile function F^{-1} (u). The generalized Tukey-Lambda distribution is defined in terms of its percent point function. Note that there are two parameterizations of this distribution in the literature.

Furthermore, the Generalization of Tukey's lambda distribution [Hastings et al., 1947; Tukey, 1960, 1962]. Tukey's lambda distribution is symmetric and is defined by its quantile function. It was introduced to efficiently redistribute uniform random variates into approximations of other distributions [Tukey, 1960; Van Dyke, 1961]. Soon after the introduction of Tukey's lambda distribution, Hogben [1963], Shapiro et al. [1968], Shapiro and Wilk [1965], and Filliben [1975] introduced the non-symmetric case in sampling studies.

In order statistics [Ozturk and Dale, 1985] and with percentiles [Karian and Dudewicz, 1999, 2000; Fournier et al., 2007; King and MacGillivray, 2007; Karian and Dudewicz, 2003]; the methods of moments [Ozturk and Dale, 1982; Gilchrist, 2000], L-moments [Gilchrist, 2000; Karvanen and Nuutinen, 2008], and trimmed L-moments [Asquith, 2007]; and the goodness-of-fit method with histograms [Su, 2005] and with maximum likelihood estimation [Su, 2007].

On the other side, stochastic methods have been introduced with various estimators such as goodness-of-fit [Lakhany and Mausser, 2000] or the starship method [King and MacGillivray, 1999]. Moreover, Shore [2007] studied the L2-norm estimator that minimizes the density function distance and the use of nonlinear regression applied to a sample of exact quantile values.

Over the years, Tukey's lambda distribution became a versatile distribution, with location, scale, and shape parameters, that could accommodate a large range of distribution forms. It gained use as a data analysis tool and was no longer restricted to approximating other distributions.

This is in contrast to the standard parameterization, where the asymmetry and steepness are described by a combination of the two tail indices. In Section 2 we review the history of the Generalized Tukey Lambda Distribution.

The FPL distribution is an extension of the four-parameter generalized lambda distribution (Ramberg & Schmeiser, 1974), itself an extension of Tukey's three-parameter lambda distribution (Tukey, 1962). This family of distributions has seen infrequent use within the statistical literature.

3.2.THE GENERALIZED TUKEY LAMBDA DISTRIBUTION

The First Generalized Tukey Lambda Distribution is done by Tukey, and the Tukey's _ family of distribution is defined by the quantile function $F^{-1}(u;\lambda_1,\lambda_2,\lambda_3,\lambda_4)$, given by Ramberg and Schmeiser has the percent point function

$$F^{-1}(u;\lambda_1,\lambda_2,\lambda_3,\lambda_4)=\lambda_1+\frac{\left[u^{\lambda_3}+(1-u)^{\lambda_4}\right]}{\lambda_2};\lambda_2\succ 0, \qquad (3.1)$$

with $\lambda_1,\lambda_2,\lambda_3 and\ \lambda_4$ denoting the location, the scale, and the two shape parameters, respectively. One drawback of this parameterization is that it does not define a valid probability distribution for certain values of the parameters. Furthermore, the regions that do not define a valid probability distribution are not simple.

For this reason, Friemer, Mudholkar, Kollia, and Lin developed an alternative parameterization, referred to as the FMLKL generalized Tukey-Lambda distribution, that has the percent point function

$$F^{-1}(u;\lambda_1,\lambda_2,\lambda_3,\lambda_4)=\lambda_1+\left(\frac{1}{\lambda_2}\right)\left[\frac{u^{\lambda_3}-1}{\lambda_3}-\frac{(1-u)^{\lambda_4}-1}{\lambda_4}\right];\lambda_2\succ 0$$

with $\lambda_1,\lambda_2,\lambda_3 and\ \lambda_4$ denoting the location, the scale, and the two shape parameters, respectively. where λ_1 is a location parameter, λ_2 an inverse scale parameter and λ_3,λ_4 are shape parameters.

As noted by Gilchrist [2000], one of the criticisms of the TLD is that its skewness is expressedin terms of both tail indices λ_3 and λ_4 in Eqs. (3.1) and (3.2). In one approach addressing this concern, a five-parameter TLD was introduced by Joiner and Rosenblatt [1971], which, expressed in the FKML parameterization.

It is easy to see that the quantile density quantile function is

$$f_{F^{-1}(u;\lambda_1,\lambda_2,\lambda_3,\lambda_4)}=\frac{\lambda_2}{u^{\lambda_3}+(1-u)^{\lambda_4}}$$

Note that $\lambda_3=0\ or\ \lambda_4=0$ results in division by zero in the above formula.

Although several definitions for the GTLD are available, we prefer the parameterization of Generalized Tukey Lambda Distribution (GTLD) is a very flexible distribution that can approximate, or is equal to exactly, many other distributions for appropriately chosen parameters.

The parameterization of the extended Tukey's lambda distributions has a long history of development. Contemporary parameterizations are those of Bachioua generalized the Tukey lambda distribution with four and five parameters .

It has also been used in meteorology in studying the fatigue lifetime prediction of materials in simulations of queue systems [Bachioua , 1999]since it is defined for all parameter value choices, with the exception that the scale parameter must be real numbers. The Generalized Tukey Lambda Distribution is usually defined in terms of its five-parameters percentile function:

$$F^{-1}(u;\lambda_1,\lambda_2,\lambda_3,\lambda_4,\lambda_5)=\lambda_1+\lambda_2^{-1}u^{\lambda_3}+\lambda_5^{-1}(1-u)^{\lambda_4};\lambda_2\neq 0,\lambda_5\neq 0$$

Where: $0\leq u\leq 1$ and λ_1 the location parameter, λ_2, $\lambda_5$the scale parameters, λ_3 related to skewness, and λ_4, related to kurtosis. The five qualitatively different shapes of the TLD: Unimodal, U-shaped, monotone and S-shaped. Excepting Corrado [2001] and Tarsitano [2004], there has been little application of the TLD to financial matters.

However, as will be covered in this chapter, the shape range of the TLD family is so large that it can accommodate almost any financial time series. Analysts can therefore work with a single distribution, rather than needing several.

The ETLD has been applied to numerous fields, including option pricing as a fast generator of financial prices [Corrado, 2001], and in fitting income data [Tarsitano, 2004], in corrosion studies [Najjar et al., 2003], in independent component analysis [Karvanen et al., 2000], and in statistical process control [Pal, 2004; Bachioua, 2006; Fournier et al., 2006]. The new extension Generalized Tukey Lambda Distribution is usually defined in terms of its five-parameters percentile function

$$\begin{aligned}&F^{-1}(u;\lambda_1,\lambda_2,\lambda_3,\lambda_4,\lambda_5,\lambda_6,\lambda_7,\lambda_8)=\\&=\lambda_1+\lambda_2^{-1}u^{\lambda_3}+\lambda_5^{-1}(1-u)^{\lambda_4}+\lambda_6^{-1}u^{\lambda_7}(1-u)^{\lambda_8};\lambda_2,\lambda_5,\lambda_6\neq 0\end{aligned}$$

This representation is denoted for Extended Generalized Tukey Lambda Distribution (RS ETLD) referring to the parameterization.

However, the probability density function (PDF) related to new formula does not specify a proper PDF for all combinations of the shape parameters λ_3 and λ_4 (see Fournier et al. (2007) for the six regions in which a proper PDF is well defined). This limitation of the RS ETLD becomes problematic especially when estimating the parameters of the ETLD.

More specifically, any (λ_3, λ_4) estimate combination that is not part of the specified six regions would not produce a valid PDF. In order to avoid this problem, Freimer et al. (1988) propose a different parameterization for the ETLD denoted Freimer-Mudholkar- Kollia-Lin Generalized Tukey Lambda Distribution.

The Generalized Tukey Lambda Distribution (GTLD) offers an alternative. The four-parameter TLD family is known for its high flexibility. It can create distributions with a large range of different shapes. It shares the heavy-tail and skewness properties of the α-stable distribution.

The more intuitive parameterization of the ETLD that expresses the location and scale parameters directly as the median and interquartile range of the distribution. The two remaining shape parameters respectively characterize the asymmetry and steepness of the distribution.

This becomes especially useful in the estimation of risk measures, where the choice of the distribution is crucial for accuracy. Furthermore, the ETLD is defined by a quantile function. Its parameters can be estimated even when its moments do not exist.

3.3.MEASUREMENT OF SKEWNESS AND KURTOSIS FOR THE TUKEY LAMBDA DISTRIBUTION

Skewness is a measure of symmetry, or more precisely, the lack of symmetry. A distribution, or data set, is symmetric if it looks the same to the left and right of the center point. Skewness is a measure of the asymmetry of a distribution. It tells us how observations are distributed around the mean. The Skewness and Kurtosis defined respectably by:

$$\begin{cases} \text{Skewness} = \sum_{i=1}^{n}\left(X_i - \overline{X}\right)^3 / \left(n\hat{\sigma}^3\right) \\ \text{Kurtosis} = \sum_{i=1}^{n}\left(X_i - \overline{X}\right)^4 / \left(n\hat{\sigma}^4\right) \end{cases}$$

If most observations lie to the left of the mean, we say that the distribution is skewed to the right, or positively skewed. This kind of distribution has a long tail on its right side. When most observations lie to the right of the mean, we say that the distribution is skewed to the left, or negatively skewed. A distribution with a negative skew has a long tail on its left side.

Skewness essentially measures the symmetry of the distribution, while kurtosis determines the heaviness of the distribution tails. Generalizations of the symmetric Tukey lambda distributions to asymmetric random variables are considered.

These distributions, defined via inverses F-1 of their distribution functions, yield simple algorithms useful in Monte Carlo studies and they are also useful in fitting data. Here a recently recommended measure of skewness is used to determine the skewness of such distributions.

During the exam, it is important to know what the relationship among the mean, the median, and the mode is if we're dealing with a skewed distribution, irrespective of whether it's skewed to the right or to the left. This skewness measure is also adapted to the measurement of the kurtosis of members of the symmetric class.

The fourth central moment is called kurtosis. Kurtosis is a measure of the peakedness of a distribution or, in other words, a measure of the concentration of results. It tells us how strongly the observations are clustered around the mean. Now let's move on to kurtosis. Variance can be said to be the second central moment. Skewness is often referred to as the third central moment. Kurtosis is a measure of whether the data are heavy-tailed or light-tailed relative to a normal distribution.

The results from the skewing methodology as well as from the formula for the rth order L-moments will be applied to well-known symmetric univariate distributions. These include the arcsine, uniform, cosine, normal, logistic, hyperbolic secant and Student's distributions, which do not have a shape parameter, as well as the quantile-based Tukey lambda distribution which has a kurtosis parameter.

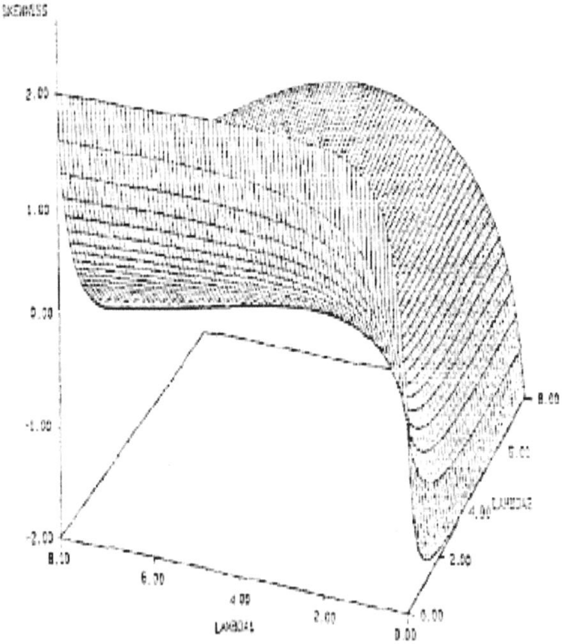

Figure(3.1): skewness and kurtosis surfaces of the lambda distribution

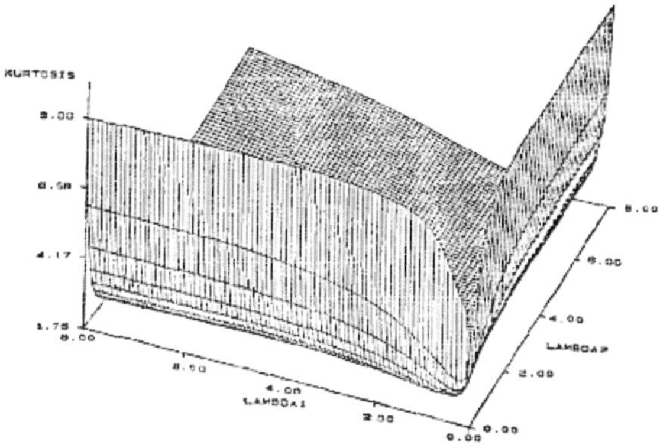

Figure(3.2): skewness and kurtosis surfaces of the lambda distribution

Clearly $f_X(x) = f_X(F^{-1}(u)) \to 0 \, or + \infty$ at the extremes of the range of variation of X, according as $f_X(x) \to 0 \quad or + \infty$, for $\lambda \succ 0$. Other modal values , if any , will correspond to solutions of the equation corresponding cases.

		$\lambda = 0{\cdot}5$		$\lambda = 1, 2$		$\lambda = 4$	
θ	ϕ	$(\alpha_3^2$,	$\alpha_4)$	$(\alpha_3^2$,	$\alpha_4)$	$(\alpha_3^2$,	$\alpha_4)$
0·5	0·5	0	1·72	0	1·50	0	1·83
0·5	1·0	0·48	2·20	0·64	2·14	0·40	2·27
0·5	2·0	0·83	3·03	1·25	3·82	0·62	2·53
0·5	5·0	1·08	3·96	1·93	7·25	1·05	3·28
0·5	10·0	1·14	4·29	2·31	9·94	1·52	5·00
1·0	1·0	0	2·08	0	1·80	0	2·45
1·0	2·0	0·36	2·45	0·57	2·40	0·10	2·53
1·0	5·0	0·64	3·08	1·18	4·20	0·35	2·24
1·0	10·0	0·73	3·38	1·52	5·78	0·77	2·91
2·0	2·0	0	2·42	0	2·14	0	3·17
2·0	5·0	0·31	2·75	0·60	2·88	−0·18	2·57
2·0	10·0	0·44	3·00	0·92	3·79	0·15	2·32
2·0	15·0	0·48	3·10	1·06	4·30	0·40	2·56
5·0	5·0	0	2.73	0	2·54	0	3·70
5·0	10·0	0·16	2·84	0·33	2·82	−0·39	3·12
5·0	15·0	0·22	2·92	0·48	3·07	−0·26	2·74
10·0	10·0	0	2·86	0	2·74	0	3·61
10·0	15·0	0·07	2·89	0·15	2·82	−0·34	3·42

Table(3.1): *Moment-ratios for*

θ	ϕ	λ	$(\alpha_3^2$,	$\alpha_4)$	θ	ϕ	λ	$(\alpha_3^2$,	$\alpha_4)$
0·5	0·5	1·44	0	1·469*	2	2	1·46	0	2·086*
0·5	1·0	1·47	0·662	2·134*	2	5	1·48	0·661	2·944
0·5	2·0	1·44	1·318	3·987	2	10	1·46	1·035	4·090
0·5	5·0	1·43	2·109	8·171	2	15	1·44	1·195	4·752
0·5	10·0	1·41	2·549	11·69	5	5	1·48	0	2·489*
1	1	1·45	0	1·753	5	10	1·51	0·380	2·836
1	2	1·44	0·604	2·401	5	15	1·48	0·551	3·155
1	5	1·45	1·299	4·549	10	10	1·49	0	2·705*
1	10	1·43	1·688	6·573	10	15	1·48	0·179	2·804*

* Minimum.

Table(3.2): Values of A giving maximum or minimum kurtosis

3.4.A SHAPE PARAMETERIZATION OF THE GTLD

It is shown how the four-parameter GTLD of Eq. (3.2) can be transformed in terms of an asymmetry and steepness parameter without adding a new variable. Moreover, the location and scale parameters are reformulated in terms of quantile statistics.

The median is used as the location parameter, and the inter quartile range is used as the scale parameter. The new parameterization brings a clearer interpretation of the parameters, whereby the asymmetry of the distribution can be more readily distinguished from the steepness of the distribution.

This is in contrast to the RS and FKML parameterizations, where the asymmetry is described by a combination of the tail indices. The new formulation allows for relaxed constraints upon the support regions, and for the existence of moments for the GTLD.

Another advantage of the new parameterization is that the estimation of the parameters for an empirical data set can be reduced to a two-step estimation problem, because the location and scale parameters can be directly estimated by their robust sample estimators.

Note that sample quintiles can always be estimated, even when moments of the distribution do not exist. lists sets of parameter values that make the TLD a good approximation for several well-known distributions. In the following section, the new parameterization is expressed in the FKML form. Consider the GTLD quantile function:

$$Q(u) = \lambda_1 + \frac{1}{\lambda_2} S(u/\lambda_3, \lambda_4),$$

Where;

$$S(u/\lambda_3,\lambda_4) = \begin{cases} \ln u - \ln(1-u) & if \quad \lambda_3 = 0, \lambda_4 = 0, \\ \ln u - \frac{1}{\lambda_4}\left[(1-u)^{\lambda_4} - 1\right] & if \quad \lambda_3 = 0, \lambda_4 \neq 0, \\ \frac{1}{\lambda_3}\left[u^{\lambda_3} - 1\right] - \ln(1-u) & if \quad \lambda_3 \neq 0, \lambda_4 \neq 0, \\ \frac{1}{\lambda_3}\left[u^{\lambda_3} - 1\right] - \frac{1}{\lambda_4}\left[(1-u)^{\lambda_4} - 1\right] & otherwise. \end{cases}$$

for $0 \prec u \succ 1$. Q is the quantile function for probabilities u ; λ_1 and λ_2 are the location and scale parameters ; and λ_3 and λ_4 are the shape parameters jointly related to the strengths of the lower and upper tails. In the limiting case u=o :

$$S(0/\lambda_3,\lambda_4) = \begin{cases} -\frac{1}{\lambda_3} & if \quad \lambda_3 \succ 0, \\ -\infty, & otherwise. \end{cases}$$

in the limiting case u=1 :

$$S(1/\lambda_3,\lambda_4) = \begin{cases} \frac{1}{\lambda_4} & if \quad \lambda_4 \succ 0, \\ \infty, & otherwise. \end{cases}$$

Plotting the density for given λ1, λ2, λ3, λ4 requires evaluating (2) and (3) for $0 \leq u \leq 1$ then plotting Q(u) against the corresponding f(Q(u)) for that value of u. When using the method of moments to find the parameters of (2), in order for there to be a finite k^{th} moment we need $\lambda_3 > -1/k$ and $\lambda_4 > -1/k$.

Since in order to find λ_1, λ_2, λ_3, λ_4 we only need the first four moments, hence we need to impose the condition $\lambda_3 > -1/4$ and $\lambda_4 > -1/4$. In addition to this Ramberg et al (1979) noted that there are certain combinations of λ_3 and λ_4 for which the distribution given by (2) is not a valid probability distribution. This undefined region is

$$1 + \lambda_3^2 < \lambda_4 < 1.8(\lambda_3^2 + 1)$$

see Karian and Dudewicz (2000). The regions 1, 2, 3, 4, 5, 6 in Figure. 3.3 are the ones for which the distribution is valid is as follows:

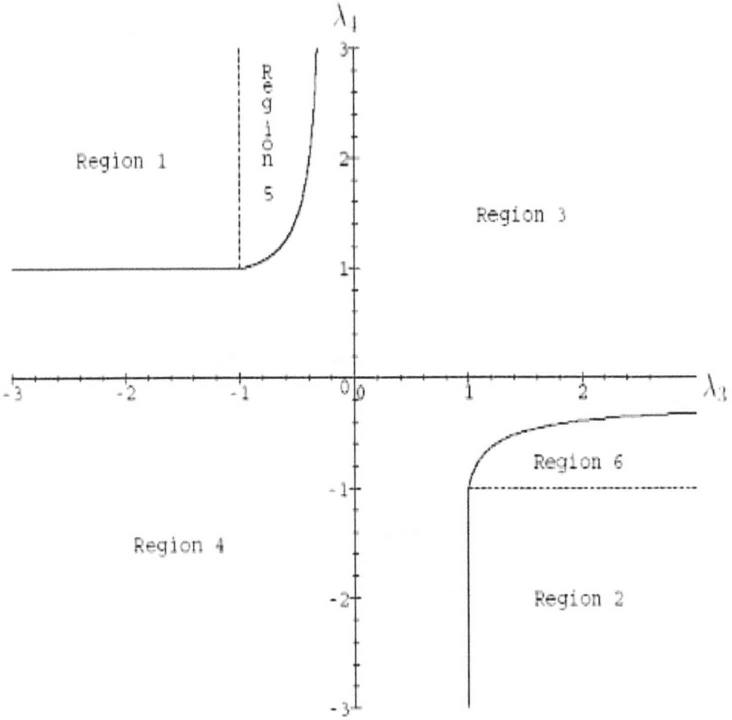

Figure(3.3): Regions 1, 2, 3, 4, 5, 6 for which the RS parameterization is valid.

3.5.SHAPE CHARACTERISTICS OF THE FMKL PARAMETERIZATION

The variety of shapes offered by this distribution includes Unimodal, U-shaped, Jshaped, S-shaped and monotone probability distribution functions, which may be symmetric and asymmetric with smooth, abrupt, truncated, long, medium or short tails. Using the relationships: x = Q(u) and F(x) = u and differentiating with respect to x, we get:

$$f(x) = f(Q(u)) = \frac{du}{dx} = \frac{du}{d\left(Q(u)\right)} = \frac{1}{\frac{d\left(Q(u)\right)}{du}},$$

Some suggestions concerning G suitable for description of the asymmetrical distributions.

This is due to the fact that the shape parameter determines also skewness, so there should be rather three linear parameters determining position, scale, and skewness together with two parameters determining the shapes of the two tails. This suggests a natural generalization and extended Tukey lambda with five to eight-parameter lambda distribution (ELDM).

This results in five class has introduced a new parameterization of the TLD, which provides an intuitive interpretation of its parameters. The median and inter quartile range are respectively synonymous with the location and scale parameters of the distribution. The shape parameters describe the asymmetry and steepness of the distribution.

Both the FMKL and RS representations are used in practice as they offer a wide variety of shapes. However, the FMKL representation is preferable as a result of its ease of use. In addition, it is well defined over all parameter values with the only restriction being that $\lambda 2$ must be positive. Also, the condition $\min(\lambda 3, \lambda 4) > -1/k$ must hold to have a finite k^{th} moment. In this study, we use the FMKL representation.

The FMKL GTLD curves are classified into five categories: Class-I family ($\lambda 3 < 1, \lambda 4 < 1$) represents Unimodal densities with continuous tails, Class-II family ($\lambda 3 > 1, \lambda 4 < 1$) represents monotone PDFs similar to the exponential distribution, Class-III family ($1 < \lambda 3 < 2, 1 < \lambda 4 < 2$)

represents U-shaped densities with truncated tails, Class-IV family ($\lambda_3 > 2$, $1 < \lambda_4 < 2$) represents S-shaped densities, and finally Class-V family ($\lambda_3 > 2$, $\lambda_4 > 2$) represents Unimodal densities with truncated tails.

So by differentiating the same formula related to the density composite functions , we find $f(x)$.

Class	Inequalities	Description
Class I	$(\lambda_3 \prec 1, \lambda_4 \prec 1)$	Unimodal densities with continuous tails. This class can be subdivided with respect to the finite or infinite slopes of the densities at the end points. Class I(a) $(\lambda_3 \prec 1, \lambda_4 \leq 1/2)$ Class I(b) $(1/2 \prec \lambda_3 \prec 1, \lambda_4 \leq 1/2)$ Class I(c) $(1/2 \prec \lambda_3 \prec 1, \lambda_4 \leq 1)$
Class II	$(\lambda_3 \succ 1, \lambda_4 \prec 1)$	Monotone PDFs similar to those of the exponential or χ^2 distributions. The left tail is truncated.
Class III	$(1/2 \prec \lambda_3 \prec 1, 1 \prec \lambda_4 \prec 2)$	U-shaped densities with both tails truncated.
Class IV	$(\lambda_3 \succ 2, 1 \prec \lambda_4 \prec 2)$	Rarely occurring S-shaped PDFs with one mode and one antimode. Both tails are truncated.
Class V	$(\lambda_3 \succ 2, \lambda_4 \succ 2)$	Unimodal PDFs with both tails truncated.

We find that the daily equity return data belongs to the Class-I family. Prior studies propose different fitting techniques for estimating the parameters of the GTLD.6 In this study, we use the maximum likelihood estimation method in the GTLDEX package of R (Su, 2007).

The computed and set equal to their GLD counterparts , and the resulting equations are solved for λ1, λ2, λ3, λ4> The complexity of the equations, to be described , forces us to seek numeric rather than closed-form solutions for λ1, λ2, λ3, λ4. Ramberg , Dudewicz , Tadikamalla and Mykytka (1979) gave tabulated solutions for λ1, λ2, λ3, λ4 for (α_3^2, α_4) restricted to the solutions available region depicted in Figure(3.3). These tables are subject to the rather server restrictions: $|\alpha_3| \prec 2$ and $1.8 + 1.7\alpha_3^2 \prec \alpha_4 \prec 9 + 1.7\alpha_3^2$. The inequality $1 + \alpha_3^2 \prec \alpha_4$, holds for all distribution , hence , it is not included here as a restriction.

This parameterization differs from previous parameterizations, where the asymmetry and steepness of the TLD is expressed in terms of both tail indices λ_3 and λ_4 of Equations. Another advantage of this new parameterization, is that the location and scale parameters can be equated to their sample estimators.

This reduces the complexity of the estimator used to fit the remaining two shape parameters. However, this new parameterization comes with the cost of having more intricate expressions for both the conditions of existence for the moments, and the conditions on the shape parameters for the different distribution shapes.

Nevertheless, evaluating these expressions remains straightforward. Moreover, the new parameterization enables the use of shape plots that can be used to represent the fitted parameters. Furthermore, value-at-risk, expected shortfall, and the tail indices can be expressed by simple formulae, dependent upon the parameters.

The TLD with unbounded support is an interesting, power-law tailed, alternative to the α-stable and the Student t-distribution for purposes of modeling financial returns. The key advantage of the TLD is it can accommodate a large range of distribution shapes.

A single distribution can therefore be used to model the data for many different asset classes; e.g., bonds, equities, and alternative instruments. This differs from current practice where different distributions must be employed, corresponding to the different distributional shapes of the data sets.

Another point, which is often overlooked in practice, is that the same distribution can be applied to different subsets of the data. For example, the distribution of the returns might completely change from year to year. The TLD would then be able to capture this distributional change.

3.6.EXERCISES

Exercise(1): If Y be the Tukey-Lambda continuous random variable with distribution defined by:

$$Y = \lambda^{-1}\left(aX^{\lambda} - (1-X)^{\lambda}\right\}, a \succ 0, \lambda \succ 0, \lambda^{-1} \le Y \le a\lambda^{-1}$$

Shaw that its nth moment about zero is given;

$$E(Y^n) = \frac{1}{\lambda^n}\sum_{j=0}^{n}(-1)^j\binom{n}{j}a^{n-j}B(\lambda(n-j)+1, \lambda_j+1),$$

Where $B(.,.)$ denotes the complete beta function> Show, in particular , that the mean and variance of Y are give by;

$$E(Y) = \frac{a-1}{\lambda(\lambda+1)} \text{ and}$$

$$Var(Y) = \frac{a^2+1}{(\lambda+1)(2\lambda+1)} - \frac{2a}{\lambda^2}\left\{B((\lambda+1),(\lambda+1)) - \frac{1}{(\lambda+1)^2}\right\}$$

Respectively.

Exercise(2): Suppose we have a sct of data from an unnown distribution. We use a Tukeys Lambda distribution with $\lambda = 0.014$ as an approximation to the density of standardized sample.

(a)-suppose the sample is actually from a $N(0,1)$distribution. What is MISE of estimated (that is , the approximated) density?.

(b)-suppose the sample is actually from a $N(\mu,\sigma^2)$distribution. What is MISE of estimated (that is , the approximated) density?.

(c)-suppose the sample is actually from a standard Cauchy distribution. What is MISE of estimated (that is , the approximated) density?.

The setting $\lambda = 0.014$ is the best fit(by matching of quantiles) of Tukey's lambda distribution to a normal distribution. The sitting $\lambda = -1$ is the best fit (by same criterion) of Tukey's Lambda distribution to a cauchy distribution.

Exercise(3):

As an exercises 1 and 2, mostly because I'd never come across the Tukey Lambda Distribution or its generalized version before, I fitted a Generalized Tukey Lambda Distribution to the residuals from the model with MARS Ratings and Interstate Status as regresses, using the GTLDEX package in R.

The Generalized Tukey Lambda Distribution has two parameterizations, one due to Ramberg & Schmeiser (RS), and the other to Friemer, Mudholkar, Kollia and Lin (FMKL).

1-What's remarkable about this distribution, which is defined by just four parameters (canonically referred to as lambda 1 through 4), is that it can approximate the Normal, Uniform, Beta, Students-t, Exponential, Weibull, Gamma, Lognormal, Pareto, F, Chi-square, Logistic, Double Exponential and Extreme Value distributions, depending on the values that its quartet of lambdas take on.

2-Find by Using maximum-likelihood estimation and the FMKL parameterization we come up with a distribution that fits the residuals fairly well - certainly at least about as well as does a Normal with the same mean and variance as the actual residuals.

CHAPTER FOUR:

ON EXTENDED TUKEY'S LAMBDA DISTRIBUTION MODELS

4.1.INTRODUCTION

In this chapter provides a new extended model with particular case five – parameter probability distribution which is an extension of generalized Tukey's lambda distribution and probability density function. Due to the highly flexible form of this new distribution, wide variety of curve shapes are possible, and thus it is useful in both fitting continuous distribution whether it is known previously, or at the availability of a random sample of data which follows unknown distribution.

This distribution differs from the so-called Generalized Lambda Distribution (GTLD) designed by Ramberg and Schmeiser (1974); see also Karian and Dudewicz (2011). Both the GTL and the GTLD distributions are two-parameter extension of the one-parameter Tukey lambda distribution of Hastings et al. (1947) and Tukey (1960), but they differ in their formulation and in the parameter values that provide approximations to asymmetric distributions. The parameter space for GTLD has gaps, which complicates MLE estimation. Moreover, while GTL and GTLD generally approximate symmetric distributions in the same way, the logistic density is a special case of GTL but, due to a technicality, can only be closely approximated by the GTLD.

This distribution is characterized by the method of estimating the parameters of this distribution also given in a variety of ways. It may be necessary to approximate many symmetric and asymmetric distributions using the proposed distribution to show the new aspects that are characterized by this distribution and confirm its importance as a qualification for other techniques to fit continuous distributions in addition to generating quasi-numbers random.

A continuous random variable $X$is usually defined by its distribution function F_X or by its probability density functionf_X.Alternatively, it can be defined by its percentile function, if exists, which is the inverse of the distribution functionF_X.This concept is given by defining the random variable U as being a function given by the following form;

$$U = F_X(X)$$

It is easy to prove that the function U, when F_X is strictly increasing function,represents a random variable that is uniformly distributed on$(0\,,1)$. Accordingly, by following the inverse transformation of this function, the random variable;

$$X = F_X^{-1}(U)$$

has $F_X$as its distribution function. Thus:

$$x = F_X^{-1}(u)\;;\; u \in (0\,,1)$$

is a well-defined function and is usually called the percentile function. It is particularly useful in Monte-Carlo simulation technique. This inverse transformation allows the probability density function$f_X$to be defined implicitly by;

$$f_X(x) = f_X[F_X^{-1}(u)] = \frac{1}{[F_X^{-1}(u)]'}\;;u \in (0\,,1)$$

Whenever$\frac{d}{du}[F_X^{-1}(u)] = [F_X^{-1}(u)]'$, this is because are in fact two important properties of all probability density function satisfies two conditions:

$$\left\{\begin{matrix} f_X(x) = f_X[F_X^{-1}(u)] \geq 0, u \in (0\,,1) \\ \int_{-\infty}^{+\infty} f_X(x)dx = \int_0^{+1} f_X[F_X^{-1}(u)]d[F_X^{-1}(u)] = 1 \end{matrix}\right\}$$

For example, let X be negative exponential distribution with parameterθ, then$f_X(x) = \theta exp\{-\theta x\}$; $\theta > 0, x \in (0\,,\infty)$, then

$F_X(x) = 1 - exp\{-\theta x\} = u$, and hence;

$x = F_X^{-1}(u) = \frac{\ln(1-u)}{\theta}$; $u \in (0\,,1)$ and$f_X(x)$can be written as:

$$f_X(x) = f_X[F_X^{-1}(u)] = \theta(1-u)\,; u \in (0\,,1)$$

Notice that $X = F_X^{-1}(U)$is in a closed form and thus it can be used for the purpose of generate observations from$X$, that is (for a given θ) by generate auniformly distributed observation from U and then set of variable of $x = \frac{\ln(1-u)}{\theta}$. However, for many considerable continuous random variables, the function $X = F_X^{-1}(U)$cannot be put in a closed form, such as normal, Weibull, extended general gamma, general beta and other random variables.

4.2. METHOD FOR CONSTRUCTING FITTING AND SIMULATIONS

In some scenarios, especially involving paired studies, symmetry of a population is a reasonable assumption. In a paired-sample experiment, a drug and a placebo might have identical distributions under a null hypothesis, in which case the difference in response variables would have a symmetric distribution.

If univariate data are assumed to be from a symmetric distribution with finite variance, then the researcher may select whether to use a t-test on the mean or a method for constructing fitting on the median since the population means and population median would be equivalent.

If the data show signs of high or low kurtosis, then the researcher might choose to use the method for constructing fitting is made after observing the data, then the false positive rates and powers are affected. Herein, this decision is based on the sample kurtosis.

The non standard values of α, the probability of the Type I error, to allow fair comparisons of simulated false-positive rates and powers between the one-sample t-test and the method for constructing fitting, when the distribution can be assumed to be symmetric with finite variance.

Three new tests are proposed, based on a mixture of the one-sample t-test and the method for constructing fitting, depending on whether the sample kurtosis is too low, too high, or either one. Nine distributions are analyzed, including the Normal, the Uniform, and seven distributions with large population kurtosis and skewness values.

The graph f_X for a given λ, we compute $x = \lambda^{-1}\left[u^{\lambda} - (1-u)^{\lambda}\right]; \lambda \neq 0$ as a value on the x-axis versus the value $\lambda^{-1}\left[u^{\lambda} - (1-u)^{\lambda}\right]; \lambda \neq 0$ on the y-axis.

Several versions of multi-parameters extensions of the above Tukey's Lambda distribution family are proposed in the literatures. The Three parameter generalization distribution was suggested in 1972, whish we call Tukeys-1. For this distribution the random variable X is specified by the percentile function :

$$X = F^{-1}(U) = T(U) = \lambda_1 + \lambda^{-1}{}_2\left[u^{\lambda_3} - (1-u)^{\lambda_3}\right]; \lambda_2 \neq 0$$

Where λ_1 represents a location parameter, λ_2 is a scale parameter, and λ_3 a shape parameter. The approximation of the function $f_X$is then given by:

$$f_X(x) = f_X(F^{-1}(u)) = T(u) = \lambda_2\lambda_3\left[u^{\lambda - 1_3} + (1-u)^{\lambda_3 - 1}\right]^{-1} ; \lambda_2 \neq 0, u \in (0,1)$$

Notice that , this distribution is symmetric a bout λ_1 , this is for the distribution of $X - \lambda_1$ and $X + \lambda_1$ are identical(since both U and $(1-U)$ have the same distribution with is uniform on $(0,1)$. Thus , this distribution is suitable for fitting the symmetrical distributions only, where methods of estimation of the parameters $\lambda_1, \lambda_2, \lambda_3$ are available by either method of moments or method of quantile .

However, for fitting asymmetrical distributions Ramberg and Schmeiser in 1974 modified Tukeys-1 distribution by suggesting the random variable X to be specified by the percentile function:

$$X = F^{-1}(U) = T(U) = \lambda_1 + \lambda^{-1}{}_2\left[u^{\lambda_3} - (1-u)^{\lambda_4}\right]; \lambda_2 \neq 0$$

Where λ_1 represents a location parameter, λ_2 is a scale parameter, λ_3 and λ_4 are shape parameters. The approximation of the function $f_X$is then given by:

$$f_X(x) = f_X(F^{-1}(u)) = T(u) = \lambda_2\left[\lambda_3 u^{\lambda - 1_3} + \lambda_4(1-u)^{\lambda_3 - 1}\right]^{-1} ; \lambda_2 \neq 0, u \in (0,1)$$

This model, which we shall call Tukeys-2 distribution, is not symmetric unless $\lambda_3 = \lambda_4$ which is distribution of Tukey's-1 as special case. Method of estimation of the parameters $\lambda_1, \lambda_2, \lambda_3$ and λ_4 are also considered , while the properties of this distribution are studied in detail by Ramberg and et al in 1979.

The four and five parameters generalization distribution was suggested in 1999 by Bachioua, which we call Tukeys-3 and Tukeys-4 respectively. For this distribution the random variable X is specified by the percentile function :

$$X = F^{-1}(U) = T(U) = \lambda_1 + \lambda^{-1}{}_2\, u^{\lambda_3} - \lambda^{-1}{}_5\, (1-u)^{\lambda_3} \;\; ;\lambda_2 \neq 0, \lambda^{-1}{}_5 \neq 0$$

$$X = F^{-1}(U) = T(U) = \lambda_1 + \lambda^{-1}{}_2\, u^{\lambda_3} - \lambda^{-1}{}_5\, (1-u)^{\lambda_4} \;\; ;\lambda_2 \neq 0, \lambda^{-1}{}_5 \neq 0$$

Where λ_1 represents a location parameter, λ_2, λ_5 are a scale parameter, λ_3 and λ_4 are a shape parameter. The approximation of the function f_X is then given by:

$$f_X(x) = f_X(F^{-1}(u)) = T(u) = \lambda_2 \lambda_3 \left[u^{\lambda-1_3} + (1-u)^{\lambda_3 - 1}\right]^{-1} \;\; ;\lambda_2 \neq 0, u \in (0,1).$$

$$f_X(x) = f_X(F^{-1}(u)) = T(u) = \left[\lambda_2 \lambda_3 u^{\lambda-1_3} + \lambda_5 \lambda_4 (1-u)^{\lambda_4 - 1}\right]^{-1} \;\; ;\lambda_2 \neq 0, \lambda_5 \neq 0, u \in (0,1).$$

4.3.THEOREMS AND RESULTS

Theorms(1): The following theorem give moments of tukey's-4 distribution.

$$\mu_j = \mu^j_{Tukey's-4} = E(X^j) = \sum_{i=0}^{j} \sum_{l=0}^{j-i} \binom{j}{i} \binom{j-i}{l} \lambda_1{}^i \lambda^l{}_2 \left(-\lambda^{-1}{}_5\right)^{j-i-l} B(\lambda_3 l + 1, \lambda_4 (j-i-l) + 1), j = 1,2,3,...$$

where, $B(p,q) = \int_0^1 x^{p-1} (1-x)^{q-1} \;\; , p \succ 0, q \succ 0.$

Proof: since,

$E(X^j) = \int_{-\infty}^{+\infty} x^{p-1} (1-x)^{q-1} dx = \int_0^1 \left[\lambda_1 + \lambda^{-1}{}_2\, u^{\lambda_3} - \lambda^{-1}{}_5 (1-u)^{\lambda_4}\right] du.$ By using the binomial expansion formula twice on $\left[\lambda_1 + \lambda^{-1}{}_2\, u^{\lambda_3} - \lambda^{-1}{}_5(1-u)^{\lambda_4}\right]^j$ and $\left[\lambda_1 + \lambda^{-1}{}_2\, u^{\lambda_3} - \lambda^{-1}{}_5(1-u)^{\lambda_4}\right]^k$ we get the result.

The above theorem provides a method of estimation the parameters of Tukey's distribution model , whenever the moments;

$$\mu_j = \mu^{j}_{Tukey} = E(X^j), j = 1,2,3,...$$

Of the original distribution of X are exist.

This becomes a tedious, laborious task. Therefore, we propose using a flexible distribution for each margin: a Generalized Tukey lambda (EGTL) distribution. This distribution allows symmetry or asymmetry and thick or thin tails. It nests a logistic distribution but also approximately nests other familiar distributions. With EGTL distributions as marginal's, the only remaining task is to choose a suitable copula. Let us therefore now examine the EGTL distribution.

The EGTL Distribution The Generalized Tukey lambda distribution was first proposed as a link function by Pregibon (1980) in the context of a generalized linear model and was analyzed in detail by Freimer et al. (1988). The random variable ϙ from the GTL distribution is given by a quantile function Q(u) for $\lambda_3 = \lambda_4$. Also this model defines another new distribution model when $\lambda_3 \neq \lambda_4$. This distribution has one parameter more than tukey's-2 distribution, hence we expect to have more curves imbedded in it. i.e. provide extra flexibility to the probability density function. Also , it is obvious that this distribution is not symmetric only if $\lambda_3 = \lambda_4$ and $\lambda_2 = \lambda_5$.

Theorms(2): The following theorem give moments of tukey's-4 distribution. The moment's estimators $\hat{\lambda}_1, \hat{\lambda}_2, \hat{\lambda}_3, \hat{\lambda}_4$ and $\hat{\lambda}_5$ of the parameters $\lambda_1, \lambda_2, \lambda_3, \lambda_4$ and λ_5 respectively of Tukey's-4 distribution are given by solving the following five non-linear system of equations:

$$\begin{cases} \mu_1 = \lambda_1 + \frac{1}{\lambda_2(\lambda_3+1)} - \frac{1}{\lambda_5(\lambda_4+1)} \\ \mu'_2 = \lambda_1 + 2\lambda_2\left[\frac{1}{\lambda_2(\lambda_3+1)} - \frac{1}{\lambda_5(\lambda_4+1)}\right] + \frac{1}{\lambda^2{}_2(2\lambda_3+1)} - \frac{1}{\lambda^2{}_5(2\lambda_4+1)} - \frac{2}{\lambda_2\lambda_5}B(\lambda_3+1,\lambda_4+1) \\ \mu'_3 = \sum_{i=0}^{3}\sum_{l}^{i-3}\binom{3}{i}\binom{3-i}{l}\lambda'^i{}_1\lambda'^l{}_2(-\lambda_5)^{3-i-l}B(\lambda_3 l+1,\lambda_4(3-i-l)+1) \\ \mu'_4 = \sum_{i=0}^{4}\sum_{l}^{i-4}\binom{4}{i}\binom{4-i}{l}\lambda'^i{}_1\lambda'^l{}_2(-\lambda_5)^{4-i-l}B(\lambda_3 l+1,\lambda_4(4-i-l)+1) \\ \mu'_5 = \sum_{i=0}^{5}\sum_{l}^{i-5}\binom{5}{i}\binom{5-i}{l}\lambda'^i{}_1\lambda'^l{}_2(-\lambda_5)^{5-i-l}B(\lambda_3 l+1,\lambda_4(5-i-l)+1) \end{cases}$$

Proof: trivial.

To solve the above system of equations it is required a numerical solution using computer. Also when the distribution of X is not specified in advance and on availability of a random sample $X_1, X_2, \lambda_3,.., X_n$ from X, we replace $\mu_j = \mu^j_{Tukey's-4} = E(X^j)$ distribution in the above system of equations by $E(X^j) = \frac{1}{n}\sum_{i=1}^{n} X_i^j, j = 1,2,3,4,5.$

4.4. DISTRIBUTION OF PERFORMANCE FUNCTION AND EXAMPLES

The distribution of a continuous random variable, such as an 1-dimensional performance function, can be approximated using several approaches including the Johnson and Pearson families of distributions, Saddlepoint approximations, and the generalized beta and lambda distributions. Usually the first four statistical moments, which define the mean, variance, skewness, and kurtosis of the distribution, respectively, are used in calculating theunknown fitting parameters in each distribution-fitting model.

Johnson distribution (Johnson , 1949) is a four-parameter model based on a transformation of the standard normal variate. It includes three distinct forms (i.e., Unbounded, Bounded, and Log-Normal) in addition to the Normal form that is a special case of the unbounded form. The main disadvantage of using the Johnson distribution is that it is often difficult to determine the four unknown parameters from the moments of the sample data (Johnson et al., 1995).

In the Pearson family of distributions, one of five different types of continuous distributions is usually used depending on the limited or unlimited range of distribution in one or both directions as well as the skewness or symmetry of the distribution. The major drawback of the Pearson family of distributions is that it can lead to unstable results near the boundaries of different regions corresponding to different distribution types (Youn et al., 2006).

Saddlepoint approximations are another powerful tool for approximating distributions that rely on the cumulant generating function (CGF) of the continuous random variable. The CGF of the random variable is approximated using the first few statistical moments. The main shortcoming of this approach is that singularities can arise during the numerical procedure (Huang and Du, 2006).

Example(1): The probability density function of general gamma distribution introduced by Agarwal and Kalla given as follows;

$$f(\lambda,n,m,r;x)=\frac{\lambda^{m-r}}{\Gamma_r(k,\lambda n)}x^{k-1}[x+n]^{-r}e^{-\lambda x};x\geq 0,r,n,m,\lambda,k\succ 0$$

$$\text{Where}\,\Gamma_r(k,n)=\int_0^{+\infty}x^{k-1}[x+n]^{-r}e^{-\lambda x}dx;\quad ,r,n,m,\lambda,k\succ 0$$

It easy to show that the moments of this distribution are given by;

$$\mu'_j=E(X^j)=\lambda^{-j}\Gamma_r(k+j,\lambda n)/\Gamma_r(k,\lambda n),j=1,2,3,4,...$$

By subsisting μ'_j; j=1,2,3,4,5 in the above system of five equations and for particular cases the table given in the next wing section using computer program, gives the fitted Tukey's-4 distribution.

Before Closed this example, we like to mention that all kinds of Tukey's lambda distribution given are function of the random variables U and $(1-U)$. We suggest another possible generalization of Tukey's Models which is worth to be studied given by the form;

$$X = F_X^{-1}(U) \equiv T(U) = aU^b - c(1-U)^d + eU^f(1-U)^g + h;$$

Where a, b, c, d, e, f, g, h are real numbers, for a special case we recommended to study this model where a=c=0.

Another generalization is also possible by defining:

$f_X(x) = f_X\left(F_X^{-1}(u)\right) = \dfrac{g(u)}{F_X^{-1}(u)}, u \in (0,1)$. Such that $F_X^{-1}(u) \succ 0$ exist, and g is a continuous on $(0,1)$ satisfying the conditions:

$\dfrac{g(u)}{F_X^{-1}(u)} \geq 0$ and $\int_0^1 g(u)du = 1$, then;

$$\int_{-\infty}^{+\infty} f_X(x)dx = \int_0^1 \frac{g(u)}{F_X^{-1}(u)} dF_X^{-1}(u) = \int_0^1 g(u)du = 1$$

Where in this case all Tukey's models become special cases when $g(u)=1 \quad \forall u \in (0,1)$ we now get a new extended representation of Tukey's lambda distribution model by the general form

$$X = F_X^{-1}(U) \equiv G(T(U)) - (T(0)) = GoT(U) - GoT(0);$$

Where $G' = g$, and $G(0) \in IR$

For Example for representation of Tukey's models is in case $GoT - 1_{(0,1)}$. For the extended general Tukey's lambda distribution represented in case $GoT(U) = aU^b - c(1-U)^d + eU^f(1-U)^g, GOT(0) = h.$

Example(2): The GLD is most easily defined by its inverse CDF, which takes a simple explicit form given as (Freimer et al., 1988);

$$X = F^{-1}(|U) = T(U) = \lambda_1 + \lambda^{-1}{}_2 \left[\frac{U^{\lambda_3} - 1}{\lambda_3} - \frac{(1-U)^{\lambda_4}}{\lambda_4} \right] ; \lambda_2 \neq 0, \lambda_3 \neq 0, \lambda_4 \neq 0$$

where λ_1 is the location parameter, λ_2 is the scale parameter, λ_3 and λ_4 define the shape characteristics ($\lambda_3 = \lambda_4$ for a symmetric distribution), and y is the CDF value (0 ≤ u ≤ 1). The PDF of GLD is defined as;

$$f_X(x) = f_X(F^{-1}(|U)) = \lambda_2 \left[\frac{1}{U^{\lambda_3 - 1} + (1-U)^{\lambda_4 -}} \right];$$

The mean (α1 = μ), variance (α2), skewness (α3), and kurtosis (α4) of GLD are defined in terms of its distribution parameters (Lakhany and Mausser 2000) as

$$\begin{cases} \alpha_1 = \mu = v_1 \\ \alpha_2 = \sigma^2 = \dfrac{v_2 - v^2{}_1}{\lambda_2^2} \\ \alpha_3 = \dfrac{v_3 - 3v_1 v_2 + 2v^2{}_1}{\left(v_2 - v^2{}_1\right)^{3/2}} \\ \alpha_3 = \dfrac{v_4 - 4v_1 v_3 + 6v^2{}_1 v_2 - 3v^4{}_1}{\left(v_2 - v^2{}_1\right)^2} \end{cases}$$

where v_1 through v_4 are functions of only λ_3 and λ_4 are defined as

$$v_1 = \frac{1}{(1+\lambda_3)\lambda_3} - \frac{1}{(1+\lambda_4)\lambda_4}$$

$$v_2 = \frac{1}{(1+2\lambda_3)\lambda_3^2} + \frac{1}{(1+2\lambda_4)\lambda_4^2} - \frac{2\beta(1+\lambda_3, 1+\lambda_4)}{\lambda_3 \lambda_4}$$

$$v_3 = \frac{1}{(1+3\lambda_3)\lambda_3^3} - \frac{1}{(1+3\lambda_4)\lambda_4^3} - \frac{3\beta(1+2\lambda_3, 1+\lambda_4)}{\lambda_3^2 \lambda_4} + \frac{3\beta(1+\lambda_3, 1+2\lambda_4)}{\lambda_3 \lambda_4^2}$$

$$v_4 = \frac{1}{(1+4\lambda_3)\lambda_3^4} + \frac{1}{(1+4\lambda_4)\lambda_4^4} - \frac{4\beta(1+3\lambda_3, 1+\lambda_4)}{\lambda_3^3 \lambda_4} + \frac{6\beta(1+2\lambda_3, 1+2\lambda_4)}{\lambda_3^2 \lambda_4^2} - \frac{4\beta(1+\lambda_3, 1+3\lambda_4)}{\lambda_3 \lambda_4^3}$$

Even though GLD is a flexible distribution that can model a wide variety of distributions, its accuracy is not guaranteed in some regions in the skewness-kurtosis plane.

Example(3): the same important cases of the general function $g(u)$ as

$$g(u)=\frac{u^{p-1}(1-u)^{q-1}}{B(p,q)};p\succ 0,q\succ 0,u\in(0,1)$$
$$g(u)=2u,u\in(0,1)$$
$$g(u)=3u^2,u\in(0,1)$$

For $g(u)=2u$, the new Tukey's representation of closed form is;

$$X=F_X^{-1}(U)\equiv G(T(U))-(T(0))=GoT(U)-GoT(0)=\left[\lambda_1+\lambda_2^{-1}U^{\lambda_3}-\lambda_5^{-1}(1-U)^{\lambda_4}\right]^2-\lambda_1^2$$

And for $g(u)=2u^2$, the new Tukey's representation of closed form is;

$$X=F_X^{-1}(U)\equiv G(T(U))-(T(0))=GoT(U)-GoT(0)=\left[\lambda_1+\lambda_2^{-1}U^{\lambda_3}-\lambda_5^{-1}(1-U)^{\lambda_4}\right]^3-\lambda_1^2$$

In the general cases $g(.)$is an integral function that satisfies the condition above . The Tukey's distribution model represent by the case $G=1_{(0,1)}$.

4.5.APPLICATION OF PERCENTILE FUNCTION OF SOME KNOWN DISTRIBUTIONS:

To evaluate the merits of Tukey's –lambda Distribution Model, is essy to fit several known distributions. The following table givesthe name of the distribution and its probability density function of the fitting distribution. The parameters are estimated by the method of moments using computer program (fortran), except Cauchy distribution where we use the percentile method. Since the moment of this distribution do not exist.

The comparison between the original distribution and the fitted distribution is so encouraging , where sometimes the maximum absolute errors between them is less then 10^{-4} which verify the goodness –of-fit test such as χ^2 goodness of fit test or kolomogorov-smirnove test.

These examples show clearly that the Tukey's-Lambda distribution model is competing the other methods of fitting distributions, such as pearson-system of distributions, Jhonson-systems, and Burr-systems Distribution models.

Uniform Distribution	$f(x;a,b)=\frac{1}{b-a};$ $b\neq a\in IR, x\in IR$	$X_{Ideal}=(b-a)U+a$
	$Case \quad a=0, b=1$	$X=0.9999U-o.425\times10^{-4}$
Normal Distribution	$N(\mu,\sigma$ $N(0,1)$	$\overset{\wedge}{\lambda}_1=0.0000, \overset{\wedge}{\lambda}_2=4.910$ $\overset{\wedge}{\lambda}_3=0.1400$ $\overset{\wedge}{\lambda}_4=1.0994, \overset{\wedge}{\lambda}_5=4.910$ $X=\overset{\wedge}{\lambda}_1+\overset{\wedge}{\lambda}_2^{-1}U^{\hat{\lambda}_3}-\overset{\wedge}{\lambda}_5^{-1}(1-U)^{\hat{\lambda}_4}$
Log Normal Distribution	$LnN(\mu,\sigma)$	$\overset{\wedge}{\lambda}_1=-0.1600, \overset{\wedge}{\lambda}_2=0.081$ $\overset{\wedge}{\lambda}_3=-0.0999$ $\overset{\wedge}{\lambda}_4=0.1409, \overset{\wedge}{\lambda}_5=4.910$ $X=\overset{\wedge}{\lambda}_1+\overset{\wedge}{\lambda}_2^{-1}U^{\hat{\lambda}_3}-\overset{\wedge}{\lambda}_5^{-1}(1-U)^{\hat{\lambda}_4}$
Cauchy Distribution	$f(x;a,b)=\frac{1}{b\pi\left[\left(\frac{x-a}{b}\right)^2+1\right]};$ $a\in IR, b\in IR^*, x\in IR$ $Case \quad a=0, b=1$	$\overset{\wedge}{\lambda}_1=0.000, \overset{\wedge}{\lambda}_2=-1.004$ $\overset{\wedge}{\lambda}_3=-0.0999$ $\overset{\wedge}{\lambda}_4=-0.0994, \overset{\wedge}{\lambda}_5=-1.004$ $X=\overset{\wedge}{\lambda}_1+\overset{\wedge}{\lambda}_2^{-1}U^{\hat{\lambda}_3}-\overset{\wedge}{\lambda}_5^{-1}(1-U)^{\hat{\lambda}_4}$
General Betta Distribution	$f(x;a,b,p,q)=\frac{(x-a)^{p-1}(x-b)^{q-1}}{B(p,q,a,b)};$ $a\in IR, b\in IR, p\in IR, q\in IR$	$\overset{\wedge}{\lambda}_1=9.059, \overset{\wedge}{\lambda}_2=0.0218$ $\overset{\wedge}{\lambda}_3=0.0305$ $\overset{\wedge}{\lambda}_4=0.0103, \overset{\wedge}{\lambda}_5=0.0218$

		$X = \hat{\lambda}_1 + \hat{\lambda}_2^{-1} U^{\hat{\lambda}_3} - \hat{\lambda}_5^{-1} (1-U)^{\hat{\lambda}_4}$
General Gamma Distribution	$f(x;k,\lambda,n,p,r) = \frac{\lambda^{kn-p} x^{p-1} [x+n]^{-r} e^{-\lambda x}}{\Gamma(k,\lambda n)}$; $k,n,r,\lambda,p \succ 0$	$\hat{\lambda}_1 = 9.051, \hat{\lambda}_2 = 0.0224$ $\hat{\lambda}_3 = 0.0309$ $\hat{\lambda}_4 = 0.019, \hat{\lambda}_5 = 0.0986$ $X = \hat{\lambda}_1 + \hat{\lambda}_2^{-1} U^{\hat{\lambda}_3} - \hat{\lambda}_5^{-1} (1-U)^{\hat{\lambda}_4}$
	$n=0, r=1, k=5, \lambda=0.5$	
	$n=0, r=1, k=10, \lambda=0.5$	$\hat{\lambda}_1 = 0.089, \hat{\lambda}_2 = 0.5008$ $\hat{\lambda}_3 = 0.0013$ $\hat{\lambda}_4 = 0.4054, \hat{\lambda}_5 = 0.05648$ $X = \hat{\lambda}_1 + \hat{\lambda}_2^{-1} U^{\hat{\lambda}_3} - \hat{\lambda}_5^{-1} (1-U)^{\hat{\lambda}_4}$
	$n=0, r=1, k=26, \lambda=0.5$	$\hat{\lambda}_1 = 46.154, \hat{\lambda}_2 = 0.0164$ $\hat{\lambda}_3 = 0.0698$ $\hat{\lambda}_4 = 0.1467, \hat{\lambda}_5 = 0.0164$ $X = \hat{\lambda}_1 + \hat{\lambda}_2^{-1} U^{\hat{\lambda}_3} - \hat{\lambda}_5^{-1} (1-U)^{\hat{\lambda}_4}$

4.6. QUANTITATIVE TECHNIQUES

There are various alternatives to the mean and median for measuring location. These alternatives were developed to address non-normal data since the mean is an optimal estimator if in fact your data are normal. Tukey and Mosteller defined two types of robustness where robustness is a lack of susceptibility to the effects of no normality.

Robustness of validity means that the confidence intervals for the population location have a 95% chance of covering the population location regardless of what the underlying distribution is.

1. Robustness of efficiency refers to high effectiveness in the face of non-normal tails. That is, confidence intervals for the population location tend to be almost as narrow as the best that could be done if we knew the true shape of the distribution.

2. The mean is an example of an estimator that is the best we can do if the underlying distribution is normal. However, it lacks robustness of validity. The alternative measures of location try to balance these two concepts of robustness. This plot shows histograms for 10,000 random numbers generated from a normal, a double exponential, a Cauchy, and a Tukey-Lambda distribution.

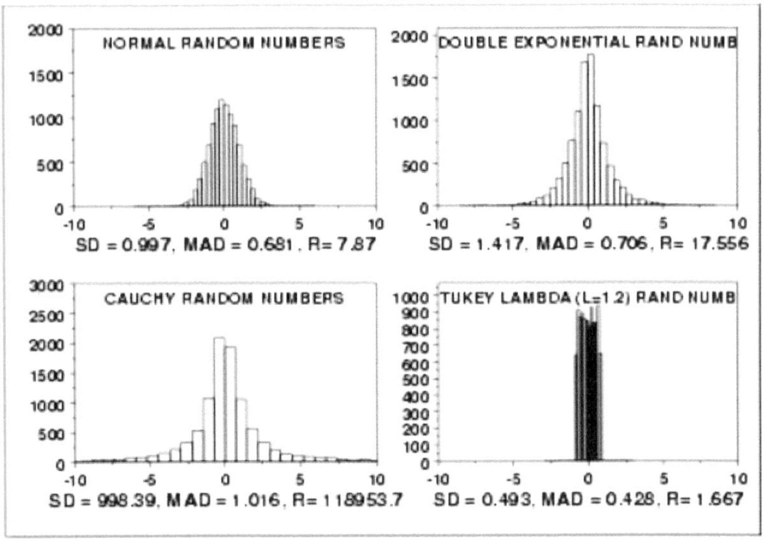

Table 5.1: Histograms of fit a extended Tukey Lambda distribution for symmetric probabilistic distribution function.

4.7.EXERCISES

Exercise(1): If T be the Tukey-Lambda continuous random variable with distribution defined by:

$$t = Q(u) \equiv Q(u;\mu,\sigma) = \mu + \sigma\left[\frac{u^{\mu-\sigma}-1}{\mu-\sigma} - \frac{(1-u)^{\mu-\sigma}-1}{\mu+\sigma}\right],$$

where u ∈ [0, 1] and μ and σ are location and scale parameters, respectively, and α and δ are shape parameters. In the following discussions, we consider the canonical form such that μ = 0 and σ = 1.

1)-Find the quantile function Q(u) translates the quantile of u into a random variable ϱ. Therefore, the CDF of this distribution F(ϱ) is defined as F(t)=u=Q $^{-1}$(t).

2)- By except for a few special values of α and δ, the function Q^{-1} does not have a closed-form expression and must therefore be evaluated numerically. Prove that the PDF f(t) is given by

$$f(t) \equiv \frac{\partial F(t)}{\partial t} = \frac{\partial Q^{-1}(t)}{\partial t} + \frac{1}{\partial Q(u)/\partial u} = \frac{1}{u^{\mu-\sigma-1} + (1-u)^{\mu-\sigma-1}},$$

2)Prove the first and second moments are given by

$$E(t) = \frac{1}{(\lambda_1+1)} + \frac{1}{(\lambda_2+1)} = \frac{-2\sigma}{(\lambda_1+1)}\frac{1}{(\lambda_2+1)},$$

$$E(t^2) = \frac{2}{(2\lambda_1+1)(\lambda_1+1)} + \frac{2}{(2\lambda_2+1)(\lambda_2+1)} - \frac{2}{\lambda_1\lambda_2}\left[B(\lambda_1+1,\lambda_2+1) + \frac{\lambda_1\lambda_2-1}{(\lambda_1+1)(\lambda_2+1)}\right],$$

Exercise(2). In particular, we adopt a five parameter version of the generalized lambda distribution

$$X(p,\lambda) = \lambda_1 + \frac{\lambda_2}{2}\left\{(1-\lambda_3)\left[\frac{p^{\lambda_4}-1}{\lambda_4}\right] - (1+\lambda_3)\left[\frac{q^{\lambda_5}-1}{\lambda_5}\right]\right\},$$

Prove that if the quantile function is a continuous and increasing function of . Here, controls, albeit not exclusively, the location of an

FPLD; the parameter acts for a multiplier to the translated quantile function and is, thus, a scale parameter; influence the shape of. The limiting cases and/or must be interpreted according to L'Hopital's rule.

Exercise(3): In particular, we adopt a five-parameter version of the generalized lambda distribution;

$$f(x) \equiv F^{-1}(u;\lambda_1,\lambda_2,\lambda_3,\lambda_4,\lambda_5) = \lambda_1 + \lambda_2^{-1}u^{\lambda_3} - \lambda_5^{-1}(1-u)^{\lambda_4} \;;\lambda_2,\lambda_5 \neq 0$$

In order to have a valid distribution , the PDF of GLD must satisfy the following conditions:

1-For all x over the allowed domain, $f(x) \geq 0$

2- $\int_{Q(0)}^{Q(1)} f(x)dx = 1$

Exercise(4): In general, we adopt a five-parameter version of the generalized lambda distribution

$$X = F^{-1}(|U) = T(U) = \lambda_1 + \lambda^{-1}{}_2 U^{\lambda_3} - \lambda^{-1}{}_5(1-U)^{\lambda_5} \;;\lambda_2 \neq 0, \lambda_5 \neq 0$$

where λ_1 is the location parameter, $\lambda_2\, and\, \lambda_5$ are the scale parameter, λ_3 and λ_4 define the shape characteristics ($\lambda_3 = \lambda_4$ for a symmetric distribution), and y is the CDF value (0 ≤ u ≤ 1).

1-Find the PDF of EGLD .

2- Find λ_3 and λ_4 that would such that λ_3 and $\lambda_4 \succ -0.25$.

$$\text{Minimize} \left(\frac{\alpha_3(\lambda_3,\lambda_4)}{\alpha_3^*} - 1\right)^2 + \left(\frac{\alpha_4(\lambda_3,\lambda_4)}{\alpha_4^*} - 1\right)^2$$

3-Find β_3 (skewness) and β_4 (Kurtosis) that would

$$\text{Minimize} \left(\frac{\alpha_3(\beta_3,\beta_4)}{\alpha_3^*} - 1\right)^2 + \left(\frac{\alpha_4(\beta_3,\beta_4)}{\alpha_4^*} - 1\right)^2$$

CHAPTER FIVE:

SOME APPLICATION OF EXTENDED TUKEY'S LAMBDA DISTRIBUTION MODELS

5.1.INTRODUCTION

In this chapter we address one of the two limitations of the Extended Generalized Lambda Distribution(EGLD) family is used to fit and modeling wide range of difference data. It is introduced by its quantile function. We introduce a more intuitive parameterization of the EGLD that expresses the location and scale parameters directly as the median and inter-quartile range of the distribution.

A simulation study performed for fitting Extended Generalized Lambda Distribution (EGLD) for financial applications, also EGLD was used for risk management and failure time distribution of component, sub systems and top-level system.

Since the general form of probability distribution can be expressed in terms of the standard distribution, the subsequent formulas are given for the standard form of the function.

Moreover, the new parameterization can be used to compare data set inconvenient asymmetry and steepness shape plot.

The class family of lambda distributions(LDs) was first proposed by Tukey in special case then it had been shown to have some useful applications.

The flexibility of Lambda Distributions as summing a wide variety of shapes has been used extensively to fit and model a wide range ,such sits applications in meteorology and modeling financial data in addition to the Monte Carlo simulations tidies.

In fact, the inverse problem of finding the probability distribution from a given moments sequence is hard as shown first by Stieltjes . The first Generalized Lambda Distribution (GLD) was presented by Tukey(1960). The Tukeys_family of distributions defined by the quantile function;

$$T(\lambda;u) \cong \begin{cases} \dfrac{\left[u^{\lambda} + (1-u)^{\lambda}\right]}{\lambda}; \lambda \neq 0, 0 \prec u \prec 1 \\ \dfrac{\log(u)}{1-u}; \lambda \to 0 \end{cases}$$

Where u is value of uniformly distributed random variable on (0,1). It can be shown that the rectangular and logistic distributions are also members of the above family of curves. So, this family has the ability to assume a wide variety of shapes including the standard distribution types of exponential, normal, , uniform, log-normal etc and also, GLD is useful and convenient in dealing with the percentile, (e.g. risk management).

Special attention is also paid to methods that allow a symmetric source distributions. Asymmetric distributions occur in important applications such as communications and biomedical signal processing. Adaptive techniques are proposed for the modeling of score functions or estimating functions. For this reason, Friemer, Mudholkar, Kollia, and all in development alternative parameterization, referred to as the FMKL generalized Tukey-Lambda distribution which has the following percent point function:

$$F^{-1}(u;\lambda_1,\lambda_2,\lambda_3,\lambda_4) = \lambda_1 + \lambda^{-1}{}_2\left[u^{\lambda_3} - (1-u)^{\lambda_4}\right]; \lambda_2 \neq 0$$

where $\lambda_1, \lambda_2, \lambda_3, \lambda_4$ denote the location, the scale, and the two shape parameters respectively.

The GLD is summarized by the inverse distribution function with four parameters. Another generalization of Tukey's Lambda family that assume a wide variety of shape is given in other form.

It offers risk management flexibility in modeling a broad range of probability distributions, and it has also been mainly used in fitting a distribution to the empirical data. The five–parameter version of generalized lambda distribution was proposed by Gilchrist (2000). The quantile function is given by;

$$F^{-1}(u;\lambda_1,\lambda_2,\lambda_3,\lambda_4,\lambda_5)=\lambda_1+\frac{\lambda_2}{2}\left[\frac{\left(u^{\lambda_4}-1\right)}{\lambda_4(1-\lambda_3)^{-1}}-\frac{(1-u)^{\lambda_4}-1}{\lambda^{-1}{}_5(1+\lambda_4)^{-1}}\right];\lambda_4,\lambda_5\neq 0$$

where $u\in(0,1),\lambda_3\in(-1,1),\ \lambda_2\succ 0$.

The latest extension of generalized lambda distribution (EGLD) is given by Karian, Dudewicz and Mc Donaldin1996. In 2006, Bachioua & Shaker presented an extended quantile function with five parameters. Also, the quantile function is given in another form:

$$F^{-1}(u;\lambda_1,\lambda_2,\lambda_3,\lambda_4,\lambda_5)=\lambda_1+\lambda_2^{-1}u^{\lambda_3}-\lambda_5^{-1}(1-u)^{\lambda_4}\ ;\lambda_2,\lambda_5\neq 0$$

where $\lambda_1,(\lambda_2,\lambda_5),(\lambda_3,\lambda_4)$ denote the location, the scale, and the two shape parameters respectively. Moreover, the author is presenting the new extended general form with eight parameters as follows:

$$\begin{aligned}F^{-1}(u;\lambda_1,\lambda_2,\lambda_3,\lambda_4,\lambda_5,,\lambda_6,\lambda_7,\lambda_8)&\cong T(u)\\&=\lambda_1+\lambda_2^{-1}u^{\lambda_3}-\lambda_5^{-1}(1-u)^{\lambda_4}+\lambda_6^{-1}u^{\lambda_7}(1-u)^{\lambda_8}\ ;\lambda_2,\lambda_5\neq 0\end{aligned}$$

In particular, we are looking for an application of this new form to approximate response time densities which had been suggested by Guatema and Van Gemund in simple case in the year2003. Indeed, response time densities and quintiles have important type for mance and is used to control quality of service metrics data, but their analytical derivation is, in general, very expensive.

Thus, in-depth discussion on the Extended Generalized Lambda family of distributions can control the range of shapes of parameters by offering the method of fitting data using the solution of system equation moments.

The generalized lambda distribution family fits the probability distributions of a wide variety of data sets, including the most important distributions en countered in the measurement applications.

Due to it sversatility, however, obtain in general formula of approximate parameters for the EGLD and a family of distributions that can take on a very wide range of shapes, very challenging with a general form. The generalized lambda distribution is proposed as a useful model for security price distributions.

When fitting the GLD to data by fixing some parameters that represent this case, the phenomena is controlled by model formula equation.

It is important that the distributions found by methods such as moment matching or least squares must be properly evaluated to determine the value of each parameter. It is evident in our examples, such distributions may have a support that does not adequately span the observed data, or they may fail to satisfy goodness-of-fit test.

Extensive simulation examples showing the reliable performance of the EGLD methods are presented in some publications related with this field. For some new estimation procedures of EGLD parameters and for further references.

The method of moments has been used to determined the parameters of the GLD that fit a given data set, negating the possibility of applying those members of the GLD family that do not possess the first equations moments and yet may provide superior fits to the data. In this chapter, we develop a method for fitting a EGLD distribution to data that is based on percentiles rather the an moments.

This approach makes a larger portion of the GLD family accessible for data fitting and also simplifies some of the computational difficulties en countered in the method of moments.

The density and the distribution function of EGLD are not available directly except in a few special cases, so it must be determined by numerical methods.

Model		λ_1	λ_2	λ_3	λ_4	λ_5	$\max_{1\le i\le 39}\lvert R(p_i)-R(p_i,\lambda)\rvert$
Beta[3,6]	GλD	0.25263	2.05295	0.10853	0.38998		0.0154
	FPLD	0.29944	0.23902	0.23451	0.29279	0.30638	0.0120
Chi-square[4]	GλD	1.16817	0.05299	0.01038	0.18126		0.0553
	FPLD	2.37635	3.99807	0.27658	0.71823	0.09292	0.0389
Gamma[2,2]	GλD	1.79032	0.07079	0.02577	0.23227		0.7068
	FPLD	2.9177	3.81181	0.26153	0.4982	0.13424	0.6090
Lognormal[0,1]	GλD	0.429	-0.21452	-0.02024	-0.1977		0.1349
	FPLD	2.62411	4.74602	-0.91609	1.89853	-0.5331	0.0517

Table 5.1: Comparison of fit a extended Tukey Lambda distribution for asymmetric probabilistic distribution function

The Pearson system and adaptively combined fixed estimating functions present some cases of EGLD in which we solve the Pearson system of differential equations. So the EGLD equations are parametric distributions and they are used to model the distributions of the mixtures.

The strength of the set parametric equations represents wide classes of distributions, including a symmetric distributions with positive and negative kurtosis, while the estimation of the parameters is still are natively simple procedure. The Extended Generalized Lambda Distribution (EGLD) is employed for modeling source distributions.

The major benefit of the EGLD is that it also takes in to a account the skewness of the distributions. We briefly review maximum likelihood approach in ICA and study how the parameters of EGLD may be estimated.

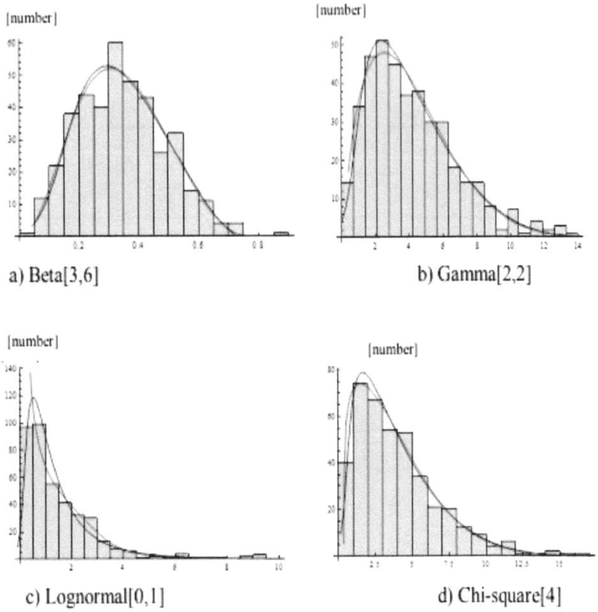

Table 5.1: Histograms of random samples and estimated functions (red curve) Comparison of fit a extended Tukey Lambda distribution for asymmetric probabilistic distribution function

.

5.2.Goodness- of- Ft Test

It is often necessary to perform multiple searches in the parameters pace before finding an acceptable distribution. We have proposed a procedure that automatically restarts the down hill simplex algorithm by generating a quasi-random initial simplex at each iteration.

The Set of pseudo sequences were used to ensure uniform coverage of the parameter space. The procedure is computationally efficient; it performs few ergood ness-of-fit tests than the starship method, for example, which makes it attractive when dealing with large samples.

Knownbestasafacilemodelforgeneratingrandomvariableswithabroad rangeof skewness and kurtosis values, the generalized EGLD has potential finantial applications, including Monte Carlo simulations, estimations of option-implied state price densities, and almost any situation requiring a flexible density shape.

The EGL Disables of distributions of ten used for parameter estimation, fitting distributions to data, has been use din studies that have included such topics or techniques as: data mining, independent component analysis. An example is used to illustrate the calculations that can be easily performed using spreadsheets.

Consequently, it is no longer necessary to assume normality and it is now possible to make direct inference on the actual at a rather than the mean of the data.

The generalized lambda distribution family fits the probability distributions of a wide variety of data sets, including the most important distribution sencountered in the measurement applications (normal, uniform, Student's, U-shaped, exponential), the shape parameters are obtained using Tukey Lambda PPCC calculator.

5.3.General Quantile Approach

Some of those defined by their quantile function are the Johnson system, with abase of normal or logistic, the Tukey lambda and its various asymmetric generalizations, with a uniform base, and Hoaglinsg-and-h, with a normal base in distribution function whose random variable is uniformly distributed on(0,1).

It can be shown that the rectangular and logistic distributions are also members of the above family. A new quantile distribution can be obtained by using the inverse function of EGLD, but it is not generally recognized how difficult this is to achieve in a symmetric distributions, whether they have one or more than one shape parameters.

The percentile of distribution function has warning limits lies between Q(0.995) and Q(0.005), and the actions limits lies between Q(0.999) and Q(0.001). The central point lie at Q(0.5).

The chart of this distribution displays the control limits. Here, a quantile control chart has been constructed using Logistic Quantile Distribution (LQD).

It may be plausible to conjecture that Closed Form (QF) expressions for Commutative distribution function (CDF) and the Probability Density Function (PDF) of the mirror-exponential do not exist. If this turns out to be the case, then the new mirror-exponential joins Tukey's lambda distribution in this some what exclusive club.

It is more common for the QF to defy analytic expression than for the CDF, and the Erlang is an example. Of course, knowledge of any one of the three basic associated functions (PDF,CDF,QF) is sufficient to represent the distribution and obtain its main parameters (albeit numerically in some cases).

5.4.Size and Power in Finite Samples

Researchers face the challenge of relevance of statistical distributions to their data Although the study of statistical modeling has made great strides in recent years, the number and variety of distributions to choose from—with all its formulas, tables, graphs, and general characteristics—still create problems for the application of Certain.

This is why many people interested in the issue of alignment ask about the future steps of the process of foresight in the future of quasi-randomized trials, which help the appropriate statistical distributions in answering these questions, which of the tens distributions should one use? And what if none of them fit well?.

Harmonization needs to focus on techniques that are successfully used in several relevant areas related to Generalized Lambda Distribution (GLD), Generalized Bootstrap (GB), and Monte Carlo (MC) simulation.

Fitting uses the tables, algorithms, and computer software needed to fit continuous probability distributions of data in a variety of conditions covering bivariate as well as univariate distributions, including situations where there are no moments.

Statistical distributions of the appropriate practitioner or theorist are required reading. It includes extensive applications illustrating methods in practice and provides evidence of key findings for those involved in theoretical development. Without it, you may be using outdated methods, wasting time, and risking incorrect results.

The random variable X has a standardized Tukey Lambda distribution if the random variable is uniform between 0 and 1, and the estimated sources are model educing EGLD. The modeling scheme provides a use full connection between practical's timed or and theoretical measure of independence.

The score function of the EGLD is use das an objective function of fitting continuous distributions that we maximize. Natural gradient and fixed-point algorithms are used to maximize the EGLD model. Also the class of GLDs is primarily used for modeling unvaried real word data.

5.5. Generating Random Numbers

The family of generalized lambda distributions is considered as a very broad family of continuous univariate probability distributions, which is why most authors, and in their forefront of investigation of this distribution, seek to thoroughly explore the relationship between the family of generalized lambda distributions with many families of commonly used distributions, the generalized lambda family of distributions. Which are used to fit the data, and generate psudo-random numbers.

Practitioners of simulation who wish to fit data to a generalized lambda distribution from the generation of many with actual datasets find the techniques useful. Tukey(1960)[27] introduced a family of random variables defined by the transformation X=F(U), where U is uniformly distributed on [0,1].

Some of its properties are described with emphasis on properties of the sample range. The rectangular and logistic distributions are members of this family and distributions corresponding to certain values of parameters give useful approximations to the normal and t distributions .

Extended Generalized Lambda Distribution EGLD is very excellent method but we still have to generate uniform distribution to start the Process of best method. We found that the Random function from Mathematica or from other software calculation packages were efficient.

It is important to test the randomness of the generator before proceeding the generation of random numbers which is done by LET subcommands. For example, the statement LETX=UNIFORMRAN DOMNUMBERSFORI=1,125generatesavariableX that contains uniform random numbers in the first25 rows.

5.6.Monte Carlo Simulations

In the Monte Carlo simulations, the fire heartily aerates are sampled from the generalized extended general Tukey distribution using the rejection sampling method. These model return the smoke layer height, X, at as specified critical time. We have arbitrarily chosen the critical time to be120s. Our objective in performing Monte Carlo simulations is to determine the accuracy of there results obtained using moment methods and the refereed do not consider issue son the computational efficiency of our Monte Carlo simulations.

In this procedure, the accuracy of the reliability estimation depends on the accuracies of the estimated statistical moments and probability distribution. The example problems in the next section are used to measure the accuracy and efficiency of the proposed approach.

Although the method proposed in this paper is straightforward and easy to implement, it has the following restrictions:

1. The method is applicable for moderate to low reliability levels. For high reliabilities, accuracy will suffer due to the tail sensitivity problem of the fitted distribution.

2. The performance function must be uni-modal (i.e., single peaked).

3. The use of DR for estimation of statistical moments requires an independent set of random variables; otherwise, the dependent variables have to be transformed into an independent set prior to using the DR procedure.

4. There is no theoretical proof for formulation invariance as altering the formulation of the performance function can potentially alter the values of EGLD parameters. Therefore, this possibility should be considered when using EGLD for distribution fitting.

5. The accuracy of reliability estimations via the proposed method diminishes as the reliability level increases, since the distribution fitting will not be as accurate along the tails of the distribution. So, we propose the use this method to be used with fairly high probabilities of failures (not smaller than 0.01).

5.7. FIT EXTENDED GENERALIZED LAMBDA DISTRIBUTION

The fitting EGLD to empirical data always requires numerical optimization; due that are only defined by the inverse of the quantile functions. It involves careful consideration of initial values for optimization, a suitable criterion to optimize and ensuring the eventual solution remains a valid probability distribution.

The Maximum Likelihood (ML) fits the mean, variances, skewness and kurtosis resemble more closely to that of empirical data, in addition to satisfying the boots trap Kolgomorov-Smirnoff (KS) goodness of fit better or equally well to starship method, whishes very flexible met hood in which the same data can be fitted by several different credible distributions, depending on the bin widths representation of the histogram.

The advantage of this method is that it allows the users to conduct sensitivity analysis by capturing the features of the empirical distribution of their interest.

The ML and starship methods are the definitive fits to the data set and will be used to obtain good fit. It provides the generally acceptable fits to empirical data using standard statistical criteria. As there is currently no single method that would work the best in every situation, the best fitted distribution chosen for a data set can be determined by fitted EGLD which best matches the moments, i.e. the mean, variance, skewness and kurtosis of the empirical data. The mean and variance of the data are of ten quite crucial.

For this reason, this criterions used rather than the alternative boot strap KS test, which is a general purpose goodness of fit procedure. This is use full as second stage selection assume times the goodness of fit result silvery similar between the different fits. At this point in time, there are node finite diagnostic tools in statistical literature to assess the quality of fit for quantile distributions such as the EGLD.

5.8. FITTING STATISTICAL DISTRIBUTIONS WITH EGLD

The extended generalized lambda distribution (EGLD) is a highly flexible distribution that has many applications in estimating the distribution of real data. Its flexibility is due to its parameters, each of which in some way determines the shape of the distribution. Since the shape of the distribution is very sensitive to changes in all s, the smallest error in estimating each parameter results in an error in the other parameters, and then results in an overall error in estimating the data distribution.

One of the most accurate and widely used methods for estimating EGLD parameters is the momentum method, which requires integrated calculations. This paper uses Gaussian quadratic to integrate EGLD equations. The results show that in terms of both the percentage index and the momentum comparison, the new method leads to a more accurate calculation of the EGLD parameters

In this part, we use Monte Carlo simulations to assess the size and power of the proposed tests; we consider well known distributions such as the normal. The generalized lambda distribution, denoted F¸, is defined in terms of the inverse of the cumulative distribution.

$$x = F^{-1}(u;\lambda_1,\lambda_2,\lambda_3,\lambda_4,\lambda_5,,\lambda_6,\lambda_7,\lambda_8) \cong T(u)$$
$$= \lambda_1 + \lambda_2^{-1}u^{\lambda_3} - \lambda_5^{-1}(1-u)^{\lambda_4} + \lambda_6^{-1}u^{\lambda_7}(1-u)^{\lambda_8} \quad ;\lambda_2,\lambda_5 \neq 0$$

This families a wide range of symmetric and a symmetric distributions. The Generalized Lambda Distribution, EGLD(λ1, λ2, λ3, λ4, λ5, λ6, λ7, λ8), is a eight-parameters family that has been used for fitting distributions to a variety of data sets.

This paper deals with the extension of the EGLD to a system, designated by EGLD, that can be used for fitting distributions to data sets that are cited in the literature as actually occurring in practice. The EGLD can achieve all possible (mean, variance, skewness, kurtosis) vectors; the

GLD could not do this.

Previous computational methods for determining λ1, λ2, λ3, λ4, λ5, λ6, λ7, λ8of the EGLD have been improved and expanded and more refined tables are provided for finding λ1, λ2, λ3, λ4, λ5, λ6, λ7, λ8. New tables for EGLD fits that are beyond the capability of the EGLD are also provided.

5.9-RELIABILITY AND RISK MANAGEMENT

To make Risk Management a regular par to factorial practice, the development and improvement of educational materials for actuaries wishing to practice in the risk management area.

The advancement of professional recognition and opportunities for actuaries in the arena of risk management. Ultimately risk management practices of actuaries will be as well documented as the practices for calculation of actuarial serves.

Any organization, that develops and produces products, has resource limits. It may be talent, capabilities, time, funding, or some combination of these.

EXAMPLE:

Let $T \geq 0$ be a continuous random variable with CDF $T(t)$ and PDF $f(t)$, the failure rate (FR) function of the EGLD is defined as

$$h(t) = lin \frac{P(T \prec t + \Delta t / T \succ t)}{\Delta t} = \frac{g(t)}{G(t)}$$

$$\Delta t \rightarrow 0$$

For the EGLD, the failure rate function $h(t)$ is

$$h(t;\sigma) = \frac{\lambda^2 \gamma}{(\lambda+1)} \frac{k^{\gamma-1}(t+1)e^{-\lambda t}}{(k^{\gamma}-1)\left[(k^{\gamma}-1)-k^{\gamma}\right]}$$

Where $k = 1 - \frac{(\lambda + \lambda t + 1)e^{-\lambda t}}{(\lambda + 1)}$

In many practical engineering circumstances, systems reliability analysis is complicated by the fact that the failure time distributions of the constituent sub systems can not be accurately modeled by standard distributions.

In this part of reliability and risk management, we present a low-cost, compositional approach based on the use of the first four statistical moments to characterize the failure time distributions of the constituent components, subsystems, and top-level system.

The approach his based on the use of Pearson distributions as an intermediate analytical vehicle, in terms of which the constituent failure time distributions are approximated.

The analysis technique is presented for-out-of-systems with identical subsystems, series systems with different subsystems, and systems exploiting stand by redundancy.

The moment-in-moment-out approach allows for the analysis of systems with arbitrary hierarchy and arbitrary (unimodal) failure time distributions, provided the sub systems are independent such that the resulting failure time can be expressed in terms of sums or order statistics.

The primary value of product reliability is in meeting the customer's expectation that the product will work as intended for sufficient time. The market rejects products that fail often, and desires products that 'just work'. Creating a reputation for a reliable product assists in increasing sales.

An extension of the value that consumers place on reliability is the willingness to pay a premium for products with high reliability. Automobiles, computers, printers, appliances and test equipment are all examples where products known for high reliability charge a premium.

The technique consistently exhibit sever good accuracy average, much less than 1 percent error at very modest computing cos. It approximates a component failure time distribution by a lambda distribution having four to eight parameters, which can be determine dif the input moments are known.

A next pension of the method is given in. Unfortunately, the lambda distribution method turns out to be less accurate, and least able in computing than the method presented in the current chapter.

In addition, the four Pearson characterization parameters can directly be expressed in the four given or measured moments, quite different from the method presented in where a table look-up or linear programming method is required to obtain these parameters. Moreover, our method enables the calculation of the failure time PDF of the composite of components of different type.

5.10.EXERCISES

Exercise(1):

The corresponding PDF frm the quantile function of the GLD is given by

$$f(x)=f(Q(u))=\lambda_2\left[\lambda_3 u^{\lambda_3-1}+\lambda_4(1-u)^{\lambda_4-1}\right]^{-1}\ ;\lambda_2\neq 0,\quad at\ x=Q(u),$$

and according the quantile density function is

$$q(u)=f(Q(u))=\lambda^{-1}{}_2\left[\lambda_3 u^{\lambda_3-1}+\lambda_4(1-u)^{\lambda_4-1}\right];\lambda_2\neq 0,$$

In order to have a valid distribution , the PDF of GLD must satisfy the following conditions:

1-For all x over the allowed domain, $f(x)\geq 0$

2- $\int_{Q(0)}^{Q(1)} f(x)dx=1$

Exercise(2): Let T be a random variable that follows the uniform distribution with support as in case (i) in exercise(1), and let X be a random variable having the Uniform R{GL} PDF. The n^{th}moment of X can be expressed in terms of the quantile function $Q_R(u)$ and is given by:

$$E_X(X^n)=\frac{1}{2}\left[\lambda_3\int_0^1 u^{\lambda_3-1}[Q_R(u)]^n du+\lambda_4\sum_{k=0}^{+\infty}(-1)^k\binom{\lambda_4-1}{k}\int_0^1 u^k[Q_R(u)]^n du\right]$$

By applying Cauchy-Schwarz inequality to $E_X(X^n)$, the nth moment of X is bounded by the upper bound provethat :

$$E_X(X^n)\leq\frac{\lambda_3}{2}\left\{\int_0^1[Q_R(u)]^n du\int_0^1 u^{2\lambda_3-2}du\right\}^{1/2}+\frac{\lambda_4}{2}\left\{\int_0^1[Q_R(u)]^{2n} du\int_0^1(1-u)^{2\lambda_4-2}du\right\}^{1/2}$$

$$=\frac{2}{2}\left[\left\{E_X(X^{2n})\right\}^{1/2}\right]\left[\left\{\frac{\lambda^2{}_3}{2\lambda_3-1}\right\}^{1/2}+\left\{\frac{\lambda^2{}_4}{2\lambda_4-1}\right\}^{1/2}\right]$$

Exercise(3): The generalized lambda distribution (GTLD) family is given in terms of its quantile or percentile function;

$$Q(u) = Q(u;\lambda_1,\lambda_2,\lambda_3,\lambda_4) = \lambda_1 + \lambda^{-1}{}_2\left[u^{\lambda_3} - (1-u)^{\lambda_4}\right];0 \le u \le 1$$

Prove that $q(u) = f(Q(u)) = \lambda_2\left[\lambda_3 u^{\lambda_3 - 1} + \lambda_4(1-u)^{\lambda_4 - 1}\right]^{-1} \ge 0,$

Exercise(4): The GLD is most easily defined by its inverse CDF, which takes a simple explicit form given as (Freimer et al.1988) . Let

$$F^{-1}(u;\lambda_1,\lambda_2,\lambda_3,\lambda_4) = \lambda_1 + \left(\frac{1}{\lambda_2}\right)\left[\frac{u^{\lambda_3}-1}{\lambda_3} - \frac{(1-u)^{\lambda_4}-1}{\lambda_4}\right];\lambda_2 \succ 0$$

1-Find density function f(x).

2-Find skwenss and kirtusess.

3-Determine $\alpha_1,\alpha_2,\alpha_3$ *and* α_4 , and solve the system of moment estimation:

$$\begin{aligned}
\hat{\alpha}_1 &= \bar{x} = \sum_{i=1}^{n} x_i/n \\
\hat{\alpha}_2 &= \hat{\sigma}^2 = \sum_{i=1}^{n}(x_i - \bar{x})^2/n \\
\hat{\alpha}_3 &= \sum_{i=1}^{n}(x_i - \bar{x})^3/(n\hat{\sigma}^3) \\
\hat{\alpha}_4 &= \sum_{i=1}^{n}(x_i - \bar{x})^4/(n\hat{\sigma}^4)
\end{aligned}$$

REFERENCES

A. Lakhany and H. Mausser., "**Estimating the Parameters of the Generalized Tukey Lambda Distribution**". Algo Research Quarterly, 3(3):47–58, 2000.

A. Ozturk and R. Dale. , "**A Study of Fitting the Generalized Tukey Lambda Distribution to Solar-radiation Data**". Journal of Applied Meteorology, 21(7):995–1004, 1982.

A. Tarsitano., "**Comparing Estimation Methods for the FPLD**". Journal of Probability and Statistics, 2010.

A. Tarsitano., "**Fitting the Generalized Tukey Lambda Distribution to Income data**". In COMPSTAT'2004 Symposium, pages 1861–1867. Physica-Verlag/Springer, 2004.

B. Dengiz. , "**The Generalized Tukey Lambda Distribution in Simulation of M/M/1 Queue Systems**". Journal of the Faculty of Engineering and Architecture of Gazi University, 3:161–171, 1988.

B. Fournier, N. Rupin, M. Bigerelle, D. Najjar, and A. Iost. , "**Application of the Generalized Tukey Lambda Distributions in a Statistical Process Control Methodology**". Journal of Process Control, 16(10):1087–1098, 2006.

C. J. Corrado. , "**Option pricing based on the Generalized Tukey Lambda Distribution**". Journal of Futures Markets, 21(3):213–236, 2001.

D. Hogben. , "Some **Properties of Tukey's Test for Non-Additivity"**. Unpublished Ph.D. thesis, Rutgers-The State University, 1963.

H. Shore., "**Comparison of Generalized Tukey Lambda Distribution (TLD) and response modeling methodology (RMM) as general platforms for distribution fitting**". Communications In Statistics-Theory and Methods, 36(13-16):2805–

2819, 2007.

J. Hastings, Cecil, F. Mosteller, J. W. Tukey, and C. P. Winsor. , "**Low Moments for Small Samples: A Comparative Study of Order Statistics**". The Annals of Mathematical Statistics, 18(3):413–426, 1947.

J. Karvanen and A. Nuutinen., "**Characterizing the Generalized Tukey Lambda Distribution by L-moments**". Computational Statistics & Data Analysis, 52(4):1971–1983, Jan. 2008.

J. Van Dyke., "**Numerical Investigation of the Random Variable $y = c(u^{\lambda} - (1 - u)^{\lambda})$**". Unpublished working paper, National Bureau of Standards, 1961.

J. W. Tukey., "**The Future of Data Analysis**". The Annals of Mathematical Statistics, 33(1):1–67,Mar. 1962.

J. W. Tukey., "**The Practical Relationship Between the Common Transformations of Percentages or Fractions and of Amounts**". Technical Report Technical Report 36, Statistical Research Group, Princeton, 1960.

J.Brian L.; Rosenblatt, Joan R., "**Some Properties of the Range in Samples From Tukey's Symmetric Lambda Distributions**", Journal of the American Statistical Association. 1971, 66 (334): 394–399.

Karvanen, Juha; N, Arto ,. "**Characterizing the Generalized Lambda Distribution by L-moments**", Computational Statistics & Data Analysis. 52 (4): 1971–1983

L. Joiner and J. R. Rosenblatt., "**Some Properties of the Range in Samples From Tukey's Symmetric Lambda Distributions**", Journal of the American Statistical Association, 66(334): 394–399, 1971.

M. Bigerelle, D. Najjar, B. Fournier, N. Rupin, and A. Iost.,

"Application of Lambda Distributions and Bootstrap Analysis to the Prediction of Fatigue Lifetime and Confidence Intervals". International Journal of Fatigue, 28(3):223–236, Mar. 2006.

M. Freimer, G. Kollia, G. Mudholkar, and C. Lin., **"A Study of the Generalized Tukey Lambda Family"**. Communications in Statistics-Theory and Methods, 17(10):3547–3567, 1988.

O.Vasicek., "**A Test for Normality Based on Sample Entropy**", Journal of the Royal Statistical Society, Series B, 1976, 38 (1): pp 54–59.

R. King and H. MacGillivray., "**Fitting the Generalized Tukey Lambda Distribution With Location and Scale-free Shape Functionals**". American Journal of Mathematical and Management Sciences, 27(3-4):441–460, 2007.

S. Su. , "**A Discretized Approach to Flexibly fit Generalized Tukey Lambda Distributions to Data**". Journal of Modern Applied Statistical Methods, 4(2):408–424, 2005.

S. Su. , "**Fitting TLD to Data via Quantile Matching Method**". In Z. Karian and E. Dudewicz, editors, Handbook of Fitting Statistical Distributions with R. CRC Press/Taylor & Francis, 2010.

S. Su. , "**Numerical Maximum Log Likelihood Estimation for Generalized Tukey Lambda Distributions**". Computational Statistics & Data Analysis, 51(8):3983–3998, May 2007.

T. Kim and H. White. On more robust estimation of skewness and kurtosis. Finance Research Letters, 1(1):56–73, Mar. 2004.

W. T. Shaw; J.McCabe., "**Monte Carlo Sampling Given a Characteristic Function: Quantile Mechanics in Momentum Space**", arXiv:0903.1592 , 2009.

Z.A., Karian and E. Dudewicz. , "**Comparison of TLD Fitting**

Methods: Superiority of Percentile Fits to Moments in L2 Norm". Journal of Iranian Statistical Society, 2(2):171–187, 2003.

Z.A., Karian and E. Dudewicz. , **"Fitting Statistical Distributions: the Generalized Tukey Lambda Distribution and Generalized Bootstrap Methods**". Chapman & Hall/CRC, 2000.

Z.A., Karian and E. J. Dudewicz. , "**Fitting the Generalized Tukey Lambda Distribution to data: a Method Based on Percentiles**". Communications in Statistics - Simulation and Computation, 28(3): 793–819, 1999.

Z.A., Karian., Dudewicz, E.J., McDonald, P., "**The Extended Generalized Lambda Distribution System for Fitting Distributions to Data: History,Completion of Theory**", tables, applications, the ``final word'' on moment fits, *Comm. in Statist.- Simul. \& Comput.* 25(3), 611-642.

Publisher: Eliva Press Global Ltd

Email: info@elivapress.com

Made in the USA
Middletown, DE
05 November 2026

36562494R00066